Paint Contractor's Manual

by

Dave Matis

Jobe H. Toole

Craftsman Book Company
6058 Corte del Cedro / P.O. Box 6500 / Carlsbad, CA 92018

Acknowledgements

The authors wish to thank the following companies and individuals for furnishing materials and information used in the preparation of various portions of this book.

Dunn-Edwards Corp. — Los Angeles, California

H.E.R.O. Manufacturing Co., Ltd. — Burnaby, British Columbia

L. Ron Hubbard — American philosopher and educator for his vital information on business technology.

Bill McNally — General Contractor, Los Angeles, California

Purdy Brush Co. — Portland, Oregon

*To all the old timers,
who gave us the benefit of their experience in the painting business*

Looking for other construction reference manuals?
Craftsman has the books to fill your needs. **Call toll-free 1-800-829-8123**
or write to Craftsman Book Company, P.O. Box 6500, Carlsbad, CA 92018 for
a **FREE CATALOG** of over 100 books, including how-to manuals,
annual cost books, and estimating software.
Visit our Web site: http://www.craftsman-book.com

Library of Congress Cataloging-in-Publication Data

Matis, Dave.
　Paint contractor's manual.

　Includes index.
　1. Painting, Industrial.　2. Paint.　3. Contractors.
I. Toole, Jobe H.　II. Title.
TT305.M37　　1985　　698'.1'068　　84-29315
ISBN 0-910460-46-9

©1985 Craftsman Book Company
Eighth printing 2003

Illustrations by Jerry O'Toole

Contents

1 Organizing Your Business **5**
   Getting Organized 7
   Company Goals 7
   Money to Meet Your Goal 8
   The Organization Board 11
   Company Meetings 13
   A Good Policy 15

2 Finding and Keeping the Right People **19**
   Buying Experience and Skills 20
   Rewards and Incentives 23
   Taxes and Insurance 24
   When to Give Pay Raises 25
   The Chain of Command 25
   Accidents 26
   When to Fire 26
   Employee Dishonesty 27

3 Putting on a Good Face **28**
   Trust is Basic 28
   Show Them You're a Professional 29
   Rely on Written Agreements 33
   Keep a Job Log 33
   Write an Operations Statement 34
   The Contractors' Image 34

4 Getting the Word Out **40**
   Know Your Company 40
   Know Your Public 42
   Writing Promotional Copy 45
   Promotional Avenues 46
   Mass Mailings 53
   Promotion and Gross Income 55

5 Introduction to Estimating **57**
   The Basics of Estimating 57
   Custom Jobs 59
   Time and Material Estimates 60
   Estimating Steps 60
   List Prices 62
   Estimating Tips 63
   Estimating Labor 65
   Estimating Stainwork 68
   Estimating Forms 69
   Reading Blueprints 78
   Estimating by the Square Foot 92
   Sample Estimate 93
   Manhour Tables 111
   Checklists 123

6 Planning the Job **135**
   Work Schedule 135
   The Role of a Foreman 136
   The Field Supervisor 139
   Job Scheduling Board 141
   Satisfying the Client 142
   Plan for Safety 145
   Plan for the Right Equipment 148

7 Preparing to Paint **150**
   Paint Selection 150
   The Right Tools 152
   The Right Brush for the Job 152
   Rollers 156
   Setting Up the Shop Area 157
   Getting the Room Ready 157
   Preparing Specific Surfaces 163
   Exterior Preparation 166

8 Doing the Painting **172**
    Painting with Flat Paint 172
    Brushing on Flat Wall Paint 173
    Rolling Flat Paint 176
    Spraying Flat Paint 178
    Painting with Enamel 179
    Painting Doors 181
    Painting Windows 187
    Staining 192
    Stripping 196
    General Painting Tips 196

9 Planning for Your Company **198**
    Financial Planning 199
    A System of Accounts 201
    Setting Up Your Accounts 202
    Payment Ledgers 205
    Collecting Your Money 205
    Job Financial Summary 208

10 Planning for Growth **209**
    Grow Gradually to Avoid Overextension .209
    Expansion is Limited by Resources 211
    Taking Gradient Steps 211
    Increasing Profits, Not Volume 212
    Find Your Level of Competency 212
    Look for Profitability 212
    Learn to Say "No" 212

Blank Forms **213**
    Paint Shopping List 213
    Common Materials Estimate 214
    Equipment Estimate 215
    Estimate Summary 216
    Payment Ledger Sheet 217
    Job Financial Summary 218

Index **219**

Chapter 1

Organizing Your Business

This book is written for painting contractors. If you've been working as a painter for several years and want to go into business for yourself — finding clients, selling the job, supervising the work, and collecting the money when it's due — this book is for you. If you've been running a paint contracting business for several years and want to compare notes with another paint contractor, this book is also for you. I'll let you look over my shoulder to see how I run my company. I expect that you'll learn enough to make the time you spend with this volume worthwhile.

Before I begin, let me explain a little about how I got into the business. Some of my experiences may sound familiar to you.

My partner and I have been in the paint contracting business for a total of 20 years. We learned the business through trial and error, asking questions, and determination. When we went into paint contracting, it seemed simple. All we needed were a couple of brushes and rollers. We learned very quickly how naive that was.

Our first job was an expensive home in an exclusive area of Malibu Beach, California. When doing custom work, you have to know a lot about preparation, color selection, paint application and, most important, getting the right price. My partner and I knew absolutely nothing about any of these subjects. So you could say we were bound to get off to a bad start. In fact, we lost our shirts, to say the least.

You might ask how we got a big custom job in the first place. Well, there's an old saying in the business world: "They saw us coming." Because we knew almost nothing about paint contracting, our first client knew he could get the job done for peanuts. He also knew we would have to do over anything he didn't like. Inexperienced paint contractors end up doing the work over and over again until the client agrees to pay up. Of course, we weren't licensed. So we couldn't sue to collect.

Naturally, the job took forever. We had to paint everything at least three times before the customer was satisfied. But there was one payoff on that job that we didn't expect. We learned more in that first month than on all the jobs we did in the next year.

Having launched myself into a career in paint contracting, I decided to find out as much about the subject as possible. I soon discovered that little has been written on the subject. I searched libraries, book stores and technical schools for anything that would help. What I found had almost no practical value to a paint contractor. So I used the only method available. I kept working and asking questions.

During those early years, my partner and I

would start many mornings over a cup of coffee at the paint store. Usually we met some old-timers there who would share their knowledge of painting with us. If we ran into a problem on a job, we would be at the paint store early the next morning trying to pin one of these guys down. This worked more often than not. But we soon learned that no one knew exactly how to solve every problem. Old-timer Joe would tell us to mix our paint one way. A few minutes later, old-timer Bob would tell us that Joe's method would never work. We should do something else. This usually left us with only one alternative. We listened to everything the old pros said. Then we went out and started experimenting until we got the right result.

We also discovered that employees working in paint stores knew less than we did about applying paint. Most paint store employees have little practical experience. Some are working as clerks because they couldn't make it as painters. Most knew enough to help the average homeowner, but quickly got lost on the finer points that concern a professional painter.

We did, however, find a couple of retired painting contractors who were working in paint stores. Without them I'd probably be in some other business today. There's no substitute for years of experience on a job. Those guys knew more tricks and time-saving methods than we could imagine.

Our business grew over the years. We went from a little two-man company operating out of the trunk of a car to a full-service painting business with 18 employees. We've done every type of paint contracting: custom residential, new custom homes, industrial, commercial (like banks and stores), tract homes, apartments, remodels, new construction and large condominium projects.

Over the years we've worked hard and learned a lot. We've worked many 18-hour days and seven-day weeks to get the job done. We've worked with general contractors, architects, homeowners, and interior designers. We've learned that every type of work requires specialized know-how — knowledge of the best and quickest way to get the job done. Using an 18-man crew to paint 185 condos is an entirely different business than doing a custom home for a designer.

Running a painting contracting business can be good work. You can make a nice living at it. And there are advantages to working for yourself. As the business grows, you're building an asset that grows in value. Of course, there are also disadvantages. It's demanding work with risks and potential problems on every job. And you have to meet and deal with the public every day. But I enjoy my work and expect that you could also.

I'm not going to explain the basics of painting here. Several books are available that describe all a homeowner needs to know to apply paint and coatings. But I am going to suggest ways a paint contractor can improve accepted application techniques. There's a difference between a Saturday afternoon craftsman who enjoys putting a coat of lacquer on a cabinet and a paint contractor who's coating hundreds of square feet of casework. I'll explain the way production painting has to be done to make a profit. And I'll also show why production painting doesn't have to mean a sacrifice in quality.

Most of this book covers the "how-to" of running a paint contracting business. It takes both good painting skill and good organizational skills to build a paint contracting company. You're not going to make it in the painting business if you don't understand production painting. And you're never going to make it as a production painter if you can't run a painting business.

Emphasis will be on what distinguishes a successful paint contracting business from a company that bumps along year after year, doing O.K. in most years, but never really becoming an established name in the business. In the years I've worked as a painter and paint contractor, I've noticed that the most successful, most profitable painting companies seem to have a lot in common. That's what I'm going to dwell on: what it takes to establish and build a successful painting business.

Before we get started, I want to warn you that I refer to "he" and "him" rather than "she" and "her" throughout this book. I do this for two reasons. First, most professional painters and owners of painting companies are male. I realize that women make good painters. And I know several women that are running successful painting companies. But men are still in the majority. The second reason is convenience. It's easier to stick to one pronoun. And I'd rather not invent a pronoun like *he/she* that would cover all the bases.

So don't think that my choice of gender is intended to exclude anyone. Every reference to the male of the species is intended to include the female. Maybe in the second edition of this book

I'll make all my pronouns female just to balance the scales.

Having covered these preliminaries, let's get down to business.

Setting up a system of organization is the subject of this chapter. Good organization is the foundation of every successful business. So that's where we'll start — with the foundation.

Getting Organized

The main difference between a freelance painter and a paint contractor is organization. It doesn't matter whether a freelance painter has his paperwork organized. He's paid for his time and craftsmanship. Organization may be irrelevant. But for everyone else in the painting business, organization is essential. Once you put that first employee on the payroll, you're running a company, and that company has to have procedures, standards and objectives. That's organization.

The foundation of every professional company is good organization. Good organization is just having a place for everything and putting everything in its place. It's deciding who does what job, what procedures to follow, and setting up guidelines for your company's success. Without organization, you have misdirected effort, confusion, neglected opportunities, waste, theft, and jobs half-completed or never started. Organization is essential in the painting business, from maintaining the company files to estimating, from painting a room to making phone calls.

How do you organize a business? Actually, it's a simple process. It starts with setting some goals. One major reason why many businesses fail in their first year is that the owner didn't have a business plan, some goals to shoot for. Reaching the goals is the reason for putting in all those long hours. If your goals aren't clear, your effort may be misdirected, wasted, or both.

Company Goals

Have you ever asked a small boy what he wants to be when he grows up? Usually you'll get an answer like, "I don't know," or "I want to be a doctor. Or maybe a plumber, like Daddy. But I'd like to be a cowboy, too. Or maybe a pilot!"

Now, that's fine for a child, but when you're talking about your business, your livelihood, your future, you should be more precise. A lot of *maybe's* and *I'm not sure's* will add up to no direction. Take my advice. Make your decision. Decide exactly where you want your business to be in ten years. There's nothing that says you can't change your goals as you go along. The important thing is to have *an express goal* as a guide.

You could start off in the painting business wanting to do top quality custom work and nothing else. After a few years, you might decide to expand and take on new construction, possibly dropping custom work altogether. That's perfectly all right, as long as you continue to set new goals and define them precisely.

Set optimistic goals. But also be realistic. Most beginners in the painting business want to have the largest, most profitable company they can imagine. That's fine. But with scope like that come problems of the same size: employee problems, cash problems, accounting problems, legal problems, and many more. Maybe you'd be more comfortable with a slightly smaller company with a few less problems.

Setting goals is even more important if you have a partner. Both of you should agree on exactly where the company is going and how it's going to get there. Bungling these initial steps — establishing company goals and ideals — is the most common cause of failure in partnerships. If you've got one or more partners, get an agreement on goals. If you don't, you'll end up with two partners in the same harness but pulling in opposite directions.

Here's an example of a company goal:

I will have eight to ten qualified painters working for me. The company will have two vans, three trucks, and all the tools and equipment necessary to do our work. We'll have a fully-equipped office with a secretary. We'll have well-established contacts in the business community and established credit where needed. There will be enough work to keep most crews busy nearly all the time. Annual volume will be $500,000 and our after-tax profit will be 5% of gross.

That's a reasonable goal. We could start working on it today. But to be sure we're on the track all the way, let's break that ten-year goal down into some intermediate goals that happen a little sooner:

After one year we should be 10% of the way to the final goal. After two years we should be 20% of the way there, and so on. After two years volume *should be $100,000, we should have two painters on the payroll, and profit after tax should be 5% of $100,000, or $5,000.* If you hit an intermediate goal sooner than expected, that's great. Simply adjust the remaining intermediate goals so they still reach the final goal at the time you established.

The final step in the goals program is to type up a neat copy of the finished product. Post it near your desk or on the back of a closet door. Study it once a week to see how you're doing and to remind yourself what your next move should be. If what you're doing is getting you closer to the goal, keep doing it. If what you're doing isn't taking you there, determine what changes need to be made that *will* get you there.

Money to Meet Your Goal
So far so good. We haven't talked about how we're going to get there yet, but at least we've established the direction and have a yardstick to measure success or failure every year along the way. Now, let's get practical. A $500,000 painting company is a pretty good-sized business. It will take some money to keep that business running. Let's figure how much.

You'll need four or five trucks, some office equipment, some specialized painting tools and equipment, and probably a small inventory of materials and supplies. The biggest investment will be in receivables and work in progress.

If your company is like many other painting companies, you'll need an investment of about $200,000 to run a yearly volume of $500,000. That probably seems like a lot of money. But a successful painting company needs that much working capital. Here's a breakdown. Allow $80,000 for receivables. At a $500,000 annual volume, you're taking in over $40,000 a month. If bills are paid about 60 days after they're sent out, that's $80,000 owed but not yet paid. Work in progress may eat up another $20,000 to $40,000 in labor and material advances before the job is finished and can be billed out. So receivables and work in progress together come to about $100,000.

You'll need roughly another $100,000 for equipment, supplies and materials. Five trucks, painting equipment, and tools will probably tie up about $75,000. Figure on spending about $25,000 for office supplies, equipment and a small inventory of painting materials.

Do you think you can get along without this $200,000, or with a lot less? I doubt that you can. I've seen some painters try. It's a constant struggle to run any business without adequate capital. And a painting business is no exception. The slightest little upset and lawsuits and lawyers become thick as flies around watermelon rotting in the August sun. Don't bet that you'll need one cent less than $200,000 in working capital to run a $500,000-a-year business.

Where are you going to get this $200,000? You can borrow some of the cash required. Banks will lend about 80% of the value of the trucks. The maximum loan is probably about $60,000. Material suppliers will bill you for materials and you can take 30 or 60 days to pay the bill. That's known as trade credit. It's like giving you a loan. But trade credit will be only $10,000 to $20,000, even for a fairly large paint contractor. That still leaves you about $130,000 short. Where's that money going to come from?

Fortunately, there's an answer. Most successful painting contractors have discovered that a profitable company will generate its own working capital. Remember our goal of a 5% profit after tax? Let's make some assumptions about the business and see how that 5% profit adds up during our ten years of growth.

We'll assume that business volume grows at $50,000 a year, reaches $500,000 at the end of ten years, and that profit averages 5% after taxes. Run that through your calculator and you'll discover that profits total just short of $140,000 for the ten years. That's the cash you need! The money's found!

But describing the process is easier than doing it. The hardest part is making that 5% after-tax profit. The next hardest part is leaving the profit in the business. Taking all the profit out of the business each year makes sustained growth impossible.

Resolve right now to earn a 5% profit after all expenses (including your salary) and taxes are paid. And then resolve to leave that profit in the business, no matter how much you would like to have a new truck or some office furniture.

It's important to keep in mind that in the example we just covered, we are looking at a business that could, ten years down the road, be a $500,000-a-year operation. Don't let the large figure of $200,000 in operating expenses throw you off. This is a long-range goal and is accomplished by an increase in volume, production and profits on a yearly basis.

The first-year goal is 10% of the $500,000 volume, or $50,000. That's a realistic goal for someone just starting out. Your operating expenses for a $50,000-a-year volume will be approximately $15,000 to $20,000 for that first year.

Remember that this is a step-by-step process. Set your long-range goals and build each year to accomplish them.

Once you've made those resolutions and have a clear goal in mind, you're ready for the next steps.

Using the Numbers
I'm sure you've seen advertisements that show a group of executives in three-piece suits seated around a conference table. At the end of the table is a large easel that holds a graph with some lines or bars or pie charts. You're supposed to infer that these executives are making an important decision based on some set of company figures.

I don't know whether decisions are made this way in large corporations. But I do know that every painting contractor needs to know what's happening in his business. And the best way to follow day-to-day activity is to keep track of the key indicators that show how the business is doing. These indicators can be like a road map that shows where you've been. Even more important, they're predictors of what's *going* to happen. Find a set of key indicators (numbers) that are easy to compile, easy to use, and easy to understand. I guarantee that these indicators will help you avoid a lot of grief and show the way to new opportunities.

The indicators you use should show how each area of your business is doing: promotion, estimates, jobs sold, production, work completed, and receipts. The system doesn't have to be complicated. In fact, the opposite is true. The key indicators should simplify your job.

I recommend that you keep track of only about six key indicators. These are explained in the following paragraphs. You may select slightly different indicators or decide to use other figures. But it would be foolish to keep track of 20 or 30 statistics in a small company. You don't want to spend any more time than necessary doing paperwork. The idea is to find the most important areas in your business and watch them like a hawk.

The indicators I follow are:

1) *Promotion:* Dollars spent on promoting company services.

2) *Estimates:* The dollar value of estimates completed and submitted to the customer.

3) *Jobs Sold:* The dollar volume of contracts signed.

4) *Production Hours:* The number of hours worked by painters.

5) *Work Completed:* The contract value of work finished in the period.

6) *Gross Income:* Dollars billed out (on invoiced work) and cash received (on cash-on-completion jobs).

Pick a cutoff time for your key indicators. Anything that happens after that day goes into the next period. You'll probably want to use a one-month period. But some figures are so important that you may want weekly tallies.

Notice that these indicators follow in a logical progression. First, you advertise. The ads produce inquiries that result in estimates. Successful bids result in signed contracts. Then the painters begin to work on the job. The job is finished and payment becomes due. Finally, payment is received.

If one indicator is falling, you can expect the indicators downstream to drop off shortly. If one area is doing well and the indicators are going up, the indicators that follow should head up in a week or two. Figure 1-1 shows how the various indicators generally correspond with each other.

Here's an example: Let's say you've been skimping on the promotion budget for several weeks. You were just too busy to do any new promotion: no letters sent out, no phone calls to contractors, and so on. What happens? You can expect fewer requests for estimates. A week or two later, the value of contracts signed will fall. The following month your painters will have less work to do. Less work will be completed. Finally, receipts will drop off.

Here's another example: You're looking at the figures for production hours. It's running at about normal. But sales and estimates are up more than

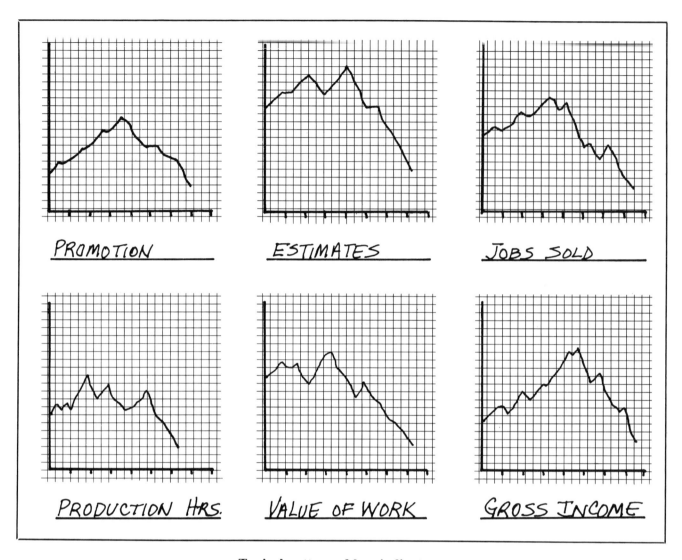

Typical pattern of key indicators
Figure 1-1

50%. You've probably developed a hefty work backlog. Some customers are waiting for work to begin. That's a bad sign if the wait is getting too long. Maybe you need to add some manpower temporarily until production is in line with sales.

Suppose the reverse is true. Sales are down but production hours are steady. It could be that your crews are stretching out the work because they suspect a layoff is coming. Closer supervision may be in order.

You can see how useful these indicators are. With a little practice, you can read them like a book. And as you develop more than one year of figures, the numbers become even more valuable. Month-to-month comparisons aren't always valid because of normal business fluctuations during the year. For example, you would expect December production and receipts to be below October production and receipts. That's normal. But if December of this year is a lot slower than December of last year, you should know why.

If you've never had your own business before, or if you're not crazy about paperwork, don't panic. Collecting numbers needed to track the key indicators doesn't take much time. And it can make your company much more efficient and profitable.

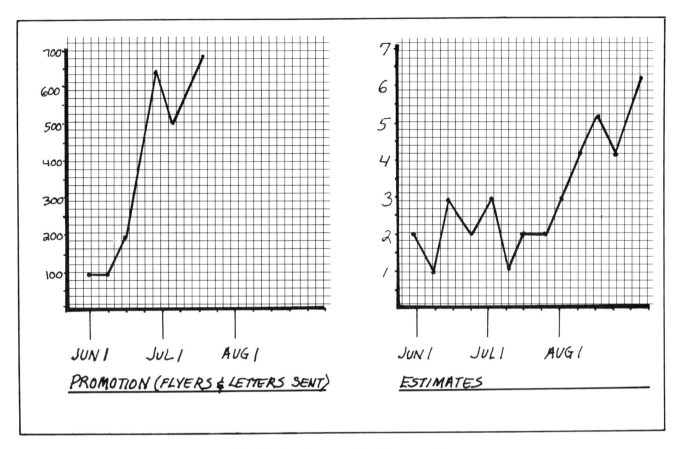

Graph showing results of promotion
Figure 1-2

The indicators help you spot problems before they happen. Sometimes the owner of a business will get so busy that he can't see the forest for the trees. I know that happens to me. The key indicators help you step back and see the big picture. They keep you in contact with the vital signs of your business. You can see what's happening right there on paper. You don't have to rely on impressions or hunches.

But be realistic about the indicators in your business. For example, just because your promotion expense suddenly jumped 50%, don't expect estimates to jump 50% the following week. The pattern of the response you get will look more like Figure 1-2. It takes people time to read those letters and respond. Not everyone who's going to respond will call in the first week or two.

Use the key indicators correctly and you'll have an excellent tool for measuring the performance of the most important areas of your business. That's a key step in getting your business organized.

The Organization Board
Look at Figure 1-3. It's an organizational chart for a painting company. Joe and Frank have identified the five major areas of responsibility in their company and either Joe or Frank has been made responsible for each. Under each area of responsibility is a list of the duties in that department. I call this chart an organization board, and I think every painting company should have one.

An organization board is an X-ray picture of your company's structure. It shows who has what job and what that job includes. In a larger company, the *Org Board* will be a very complicated diagram with lots of sub-departments and functions listed under each major division. In a smaller company, the board could be as simple as our example. The only important thing is that it makes clear who does what. It should show at a glance every significant function in your company and identify who has responsibility for that task. If you're going to be productive and show profits year after year, your organization has to run smoothly. An Org Board is designed to help it do just that.

Start your Org Board on a piece of posterboard. Keep it neat. Leave plenty of room for expansion in the tasks listed. Post the board in your office. If you don't have an office, make the Org Board small enough to carry on a clipboard or in a briefcase.

The nature of an Org Board is that it keeps growing and getting more specific. Every time there's a problem in an area that isn't listed on the board, add responsibility for that task under someone's name. If there's a question on who has responsibility for some function, change the Org Board so it answers the question. When someone new is hired, he or she should be added to the board. The board is never complete. It just keeps getting better and better in defining who does what in your company.

If you're running a one-man company, making an Org Board will identify the range of tasks that have to be done. The Org Board will help you divide your time among all the tasks. Some time has to be reserved for each task each week so that all bases are covered.

Many small painting companies neglect promotion, let bank statements accumulate unopened for several months, or fail to complete estimates because no one took charge of getting the work done. The Org Board will solve problems like that. It places responsibility clearly on some individual and makes that delegation clear to everyone in the company.

Getting Things Done

If you have the opportunity, watch a successful, highly-productive person work. As likely as not, you'll notice something about the way he tackles each problem. First, his method is probably both organized and efficient. Work follows a logical se-

JOE'S PAINTING COMPANY

MANAGEMENT	PROMOTION	FINANCES	JOBS	ESTIMATING
Joe	Joe	Frank	Frank	Joe
PAPERWORK MAKES SURE LICENSE IS UP TO DATE INSURANCE COLLECTS MONEY DOES STATISTICS	GETS FLYERS MADE MAKES SURE LETTERS OR FLYERS ARE SENT OUT RUNS ADS IN NEWSPAPERS, PHONE BOOK, ETC.	PAYS BILLS DOES PAYROLL TAXES KEEPS BOOKS	IN CHARGE OF ACTUAL WORK HIRES PAINTERS BUYS MATERIALS SCHEDULES JOBS	DOES ESTIMATES SETS APPOINTMENTS

Typical organization board
Figure 1-3

quence. There's little wasted motion or idle time. But notice something else. He probably keeps working on each task until no further action is possible. And that's the key, *completing work on each task before starting the next,* even if there isn't enough time to finish all the tasks.

What does this mean to you? I'll explain it this way. Imagine that you're at your desk and have an hour to work on an accumulated pile of correspondence, bills, notes, advertisements and phone calls that have to be returned. There isn't time to finish everything. What's your way of handling this problem?

A less-organized person would pick through the pile, pulling out something here or there that seemed interesting, working more or less at random and finishing work on little or nothing. If that's the way you usually tackle a pile of accumulated mail, there's a better way.

Let me suggest the way it should be done. First, do the easy part. Discard or file everything that doesn't need any further action on your part. Throw out the advertisements, file the receipts, sort out what has to be given to others so it can be passed on to them later. Just doing that should reduce the pile by half. Notice that you've completed all that can be done on each item discarded, filed or collected for others.

Next, set some priorities. There's only an hour available and five minutes is gone already. Set aside what can wait until more time is available. That may reduce the pile by half again. You're probably left with a small pile that needs your immediate attention. That's the place to concentrate your effort.

Work on each problem left in the pile until work is finished or nothing further can be done.

Review each invoice for accuracy. If correct, write the check, put it in an envelope with a stamp and put the envelope where you'll remember to take it to a mail box. Then file your copy of the invoice. That finishes it.

Answer phone inquiries one at a time. Return the call, answer the question or make the appointment as appropriate. When you hang up, note the time and date in your appointment book or send a confirming letter or quote immediately. File a copy. That finishes it.

If you're reviewing the monthly bank statement, scan the checks, find the total of outstanding checks and deposits, reconcile the statement, sort the checks into numerical order and file them. That finishes it.

Keep going like this, finishing as many items as possible, clearing them completely off your desk and into a file, the trash can, or a pile that you're going to give to someone else. Finishing a limited number of tasks completely is always better than working a little on all tasks. Finishing part of any job is inefficient. Time is wasted whenever you look at something and decide to do nothing or leave it half completed. To get things done, adopt this rule: *If you start it, finish it.*

This rule doesn't apply just to office work. It's true of all activities throughout the day. Starting something you don't complete leaves a little bit of your attention stuck there, whether you're conscious of it or not. Do this several times a day, day after day, and you've accumulated piles of distractions everywhere you turn. That makes reaching goals more and more difficult.

Painting is usually more efficient if you finish each part of the job before going on to the next. Assume that the job is to paint one room. First, drop it out completely. Cover everything. Use masking paper where necessary. Second, prep the room. Dig out all the cracks. Fill all the holes. Prime all the raw wood. Do your finish sanding. Dust everything. Third, paint. I know that drying times, some primers, and use of scaffolding make this impossible sometimes. But when possible, it's more efficient to complete what you start before going on.

If you aren't using this system now, try it. And encourage employees to do the same. You'll notice the improved productivity.

Company Meetings
Once a painting company has more than two or three employees, company meetings will prevent problems, resolve disputes and improve coordination. These meetings could be weekly or monthly. They could even be held only as needed. How often isn't important. What is important is that you provide some official forum for the exchange of information.

In a company with no more than six or eight employees, you probably want everyone on the payroll to be present. In a larger company, only the department heads and key field supervisors would be invited.

Keep in mind that meeting time is nonproductive

time. No work is getting done. There may not even be anyone available to answer the telephone. That's why you want to keep meetings as brief as possible. I've found that meetings held at 3:00 Friday afternoon or at 7:00 Monday morning tend to be brief and more to the point. These hours are generally less productive anyway, so we lose less productive work.

Your preparation for the meeting is important. Getting ready for a meeting forces you to sit down and take a look at your business from an executive point of view. After a hard week of painting, estimating, making phone calls, and handling customers, you need to review results, evaluate problems, and plan for the future. As the owner or partner in a business, it's your responsibility to give the company direction and momentum. You do this by reviewing company goals and evaluating progress toward those goals.

As part of your preparation, make a few notes on the topics to be covered. Something like the list in Figure 1-4 may be enough. Be sure to include a time when the floor is open for anyone to bring up any company problem. In a larger company, you may want to give all the participants copies of the agenda so they can follow what's been covered and what's coming up.

You, as the boss, call the meeting and act as chairman, president and judge. If you have one or more partners, these responsibilities are shared according to your ownership interest. In a partnership, the partners should meet privately before the meeting to agree on an agenda. You may want to meet again privately after the full meeting to reach joint decisions on problems that have yet to be resolved.

Some topics will be on the agenda at most meetings. For example, you'll want to review changes in the key business indicators (promotion, estimates, sales, production, work completed and receipts). The Org Board should be there to review and change if necessary.

One prime purpose of company meetings is to resolve problems that require joint action or a decision by the boss. Usually you'll want to cover the most important problems that have come up since the last meeting. But keep in mind that this isn't the place to administer reprimands. That should be done privately. Neither is it the place to resolve problems that concern only one or two employees. Why waste the time of those that aren't involved?

```
WEEKLY MEETING CHECKLIST
1. CURRENT STATUS OF ALL JOBS
2. UPCOMING JOBS
3. EQUIPMENT & MATERIAL NEEDED
4. REPAIRS NEEDED
5. PROBLEMS ON JOBS
6. EXTRA HELP NEEDED?
7. FINANCES
8. PROMOTIONAL ACTIVITIES
9. OPEN FORUM
```

Agenda for company meeting
Figure 1-4

Use the meeting to dispense information that everyone should know, to get ideas from everyone concerned with a problem, to coordinate the effort of all when coordination is needed, and to form a consensus on how to make the company run better.

One advantage of a company meeting is that it brings together people with different areas of responsibility and different perspectives. These people see things in a different light. Get the benefit of these perspectives. The meeting isn't just a place for the boss to pass out information. It's also a good time for the boss to learn about what's going on in the company. Make the best use of this opportunity.

No meeting should end without a concrete assignment of tasks and a memorandum of what's decided. Make notes yourself or have someone else make notes on decisions, who is to do what and when it's to be completed. Before the meeting is adjourned, read back the list of decisions, assignments and deadlines. That makes misunderstandings less likely and helps guarantee compliance by everyone concerned.

Wearing Many Hats

In any business, but especially in a small business, one person has to handle many jobs. When you finish an estimate and start on the week's bookkeeping, you're switching roles. Take off your estimator's hat and put on your accountant's hat. On most days you'll switch hats several times, from estimator to salesman to manager to painter. You may not even be aware that you're switching roles each time it happens. All of it is just your job.

In a one-man company, the definition of each job is less important. The one man has to do it all. But as a company grows, areas of responsibility have to be defined more precisely. Otherwise effort is duplicated, two people or departments will be working at cross-purposes, and some tasks will be neglected.

When you have several people on the payroll, labeling hats and defining jobs is important. Exactly what does the estimator do? What are his responsibilities? What policies does he follow? The same is true for every hat in the company. If hats are labeled and defined correctly, there should be little or no duplicated effort, less conflict, and fewer tasks neglected.

In a small painting company, job definitions don't have to be in writing. Your Org Board may be the only job definition needed. But when questions begin to come up about who should be doing what, it's time to have descriptions for the key positions. A job description makes judging performance easier, clarifies the tasks each person should handle, and simplifies the training of new employees.

Here's an example. You've been wearing all the hats in your company for two years. You know each job inside and out. But now volume is so heavy that you just can't do it all. You're ready to hire someone or promote someone to fill the position of field supervisor. The person you hire probably has experience as a supervisor. But it's unlikely that he's ready to step into your shoes and do it your way right from the start. A good job description will smooth the transition period and make the new supervisor a productive team member in the shortest time possible. He'll know exactly what's expected, the routines to follow, the people to contact, and so on.

If the person you hire has never been a supervisor before, a clear job description will shorten the training period considerably. A good job description will also eliminate many excuses, including the old I-thought-Frank-was-going-to-do-that routine.

Even in a large company, everyone needs a job description. The description doesn't have to be more than a page or two long. But it should be detailed enough to leave little room for misunderstanding. And it has to cover everything required of that job. Be sure to include in the description some standards for evaluating the employee's performance. Everyone should have a set of goals for his work, just like you have goals for the company.

How do you decide what each description should include? Easy! Just keep track of everything the person holding that job does and forgets to do for a month or two.

Figure 1-5 shows a sample job description.

A Good Policy

Every company has policies and procedures. They can be informal understandings that are never written down, or they can be impressive bound volumes. I'm going to argue that written policies are best. You don't have to go overboard, but typed pages available to everyone make it easier to enforce rules. Everyone knows exactly what the rules are. Written rules make it easier for new employees. And the rules are impartial because they're down on paper before some infraction brings a specific person into the picture.

How do you establish company policy? Easy! Policy is what has worked in the past and what you expect to work in the future. Policy is different from an Org Board and isn't like a job description. It applies to everyone in each category. It's a set of rules and regulations for those employees. Figure 1-6 is a personnel policy for office employees in a painting company.

Business, like life, is a learning experience. Early in life you made it a policy to keep your hand off hot burners. Making that mistake once is enough. It's not profitable. It doesn't lead to a positive result. It hurts. Policy statements are written to keep the same mistakes from happening over and over again. Hiring a painter with no experience to do fine custom work is not a good policy. Making that mistake once is enough.

You can make up a policy statement to cover any subject, from the smallest detail to the most obvious task. When something goes wrong because someone didn't know what to do or did the wrong thing, make a note of what should have been done.

The supervisor is responsible for:

- Follow-up on all completed jobs

- Continual supervision of all work in progress

- Preparation for all new jobs

Completed Work:
1) Make a final check on all completed work.

2) Respond to any customer complaints, and make sure all touch-up work is done to customer satisfaction.

Work in Progress:
1) Order all materials. Make sure the correct materials are delivered to the job site.

2) If there's a delay due to back-ordered materials, be sure your crew is back on the job as soon as the materials are available.

3) Make sure the job is done in proper sequence and using the correct procedures.

4) Be sure you have the right craftsmen for each particular job. If the crew you hired is not suited to the job, find one that is.

5) Check personally on the progress of the job at least twice a day.

6) When the customer has questions about the progress of the job, be sure all inquiries are directed to you and handled by you, rather than by members of your crew.

7) When the customer has questions about money or contracts, direct these inquiries to the owner of the painting company.

8) Report daily to the owner of the painting company. Let him know the status of each job.

Future Work:
1) Record new jobs on a scheduling board.

2) Review each job contract with the owner of the painting company. Determine exactly what is to be done on each job.

3) Make any special arrangements that are necessary. These might include: ordering materials in advance or juggling the schedule to make sure that the right crew is available for the job.

Supervisor's job description
Figure 1-5

Acme Painting Company - Personnel Policy for Painters

Hours: 7:00 a.m. to 3:30 p.m. One-half hour off for lunch. Two 10-minute breaks; one in the morning and one in the afternoon. Work stops at 3:15 p.m. Cleanup is from 3:15 p.m. to 3:30 p.m.

Tools: All painters are required to have the following tools:

- Two spackle blades
- One 3" vinyl brush, flat edge
- One 2" vinyl brush, angle edge
- One 3" bristle brush, flat edge
- One 2" bristle brush, angle edge
- One 4" duster
- Two screwdrivers

Days: Monday through Friday. If Saturday work is required, you'll be notified in advance.

Transportation: Each painter must provide his own transportation.

Appearance: Wear a reasonably clean set of whites. Change frequently. Torn or grimy clothing isn't allowed.

Alcohol and illegal drugs: Use of these items on the job will result in dismissal. *This is the only warning you will receive.*

Theft: Stealing company property or customer property will result in dismissal. *No exceptions.*

Payday: The work week ends on Thursday at 3:30 p.m. Time cards are delivered to the office on Thursday afternoon. Paychecks are issued Friday afternoon at the office or at the job site.

Pay adjustments: Raises result from increased productivity and responsibility. No increases are made due to length of time with the company.

Personnel policy for painters
Figure 1-6

When there's a dispute about some personnel matter, make a note of how the problem was resolved. When you have enough notes on a particular topic to fill a page, there's the first draft of your policy statement.

Give a copy of each company policy to all employees concerned. Keep everyone aware that the policy is still in effect by distributing another copy every three or four months. Give every prospective employee a copy of policy statements before you offer a job. A new employee will follow his own policies or those of a previous employer if he doesn't know your policies. Following company policies should be a condition for continued employment.

More to Come on Organization

That's as far as I'm going now on the subject of organization. True, we haven't covered everything needed to get your company organized. But we've hit most of the important high spots. There's more that you need to know. But the remainder fits best in the chapters that follow.

For now, be satisfied to set some goals, compile and review key business indicators, have an Org Board, job descriptions, and policy statements, hold effective meetings, and make it company practice to finish each task before starting the next. If you do that, you're already better organized than most painting companies.

Chapter 2

Finding and Keeping the Right People

No painting company is any better than the employees on the payroll. That makes hiring one of the most important tasks for any paint contractor.

Sure, you can do a lot to improve employee skills once they're on the payroll. You can provide good equipment that helps turn even an average painter into an exceptionally productive craftsman. You can lay down rules and policies that put every employee permanently on his or her best behavior. You can provide incentives and rewards that promote high productivity. But there's no substitute for starting out with the right people. That's the subject of this chapter, finding and keeping qualified, motivated, skilled employees.

How do you find good help? To answer that question, start by understanding the painting profession and painters.

For many years, some general contractors and a portion of the general public have assumed that painters are drunks or transients or both. You've probably heard more than one joke about an alcoholic painter. I'll admit that the painting profession has its share of misfits and losers. But there are losers, alcoholics, and misfits working as judges, doctors, teachers and cops, too. Most of the painters I know are sober, industrious, conscientious craftsmen.

Many painters do move from job to job regularly. So do carpenters, masons, electricians and plumbers. That's the nature of the construction business. Building has always been transitory work. When times are good, there are plenty of jobs. The pay is good. Everyone who knows a construction trade can get a job. When work is slow, you'll find all types of construction workers drifting from job to job. Many leave the construction industry entirely. They have to. There's no other choice.

If anything, painting offers more permanent employment than most construction trades. A building goes up only once. But it will be painted many times before reaching its normal life expec-

19

tancy. Painters will still be repainting many years after construction is complete. That gives painters better employment prospects than their brother carpenters, plumbers and electricians.

As a painting contractor, you should always be on the lookout for competent, conscientious, skilled painters. Even when you have enough good painters, keep a file of the names, addresses, and telephone numbers of painters who are available for work. If you land a large job or fall behind on scheduled work, you may need extra hands for a week or two.

Buying Experience and Skills

It's not easy to decide who is and who isn't going to be a valuable employee on the basis of a ten-minute interview. Of course, it's easy to find out how much experience someone has. Just ask him. But experience may not be the best indicator of a painter's value to your company. I've met painters with five years of experience who still haven't picked up what an apprentice should learn the first ten days on a job. Your job interview should include questions about how the prospective employee would handle specific painting problems. Third-degree interrogation isn't necessary. Just start shooting the breeze about work the applicant has done. You'll learn plenty.

If an applicant claims ten years of experience, find out exactly what *kind* of experience. Has he been doing custom work exclusively? Can he spray lacquer? Can he do stain work? If so, what kind? Does he know how birch, ash and oak will react when stained?

Anyone who considers himself a journeyman should be able to handle fine custom work, average re-do work and any new construction. He should also have some skill with staining. You can put a qualified journeyman on nearly any type of job and know that he will get the job done with a minimum of problems.

There's a big difference between painting ten new condos and doing custom work in a law office. An experienced journeyman should be capable of doing both, but most painters handle some types of work better than others. During the interview, find out if your prospect is comfortable with the kind of work your company handles.

An employment application form saves a lot of time. Figure 2-1 is modeled after the application form I use. Be sure to get the name and phone number of the applicant's last three or four supervisors. After the interview, call these supervisors. Most will be happy to answer any question you have about the applicant.

Before offering anyone a job, show him a copy of your company policy statements. Tell him what you expect of someone in the position he's being considered for. Don't leave it up to the employee to figure it out for himself. If you explain what you expect before hiring, you're more likely to get the performance you desire.

The real test, of course, comes on the job. Put the new employee on a crew with your most experienced supervisor for the first week or two. This is the probation period. You'll know in short order if the new man has what you're looking for.

Some painters are highly productive on one kind of job and have a lot of trouble with another kind. It isn't just experience, either. Painters are human beings and all people are different. You may find that a new employee is a real hotshot with a spray gun but below average on enamel brushwork. Take advantage of his strength with the gun. Have another painter pick up the slack on enamel work. A good man with a spray gun can more than carry his weight in most painting companies.

You won't find many painters who have mastered all the skills needed in painting. Any that have are probably running their own businesses. Be satisfied with painters who show skills in enough areas and are reliable.

Look for More than Painting Skill

Let's assume that you've interviewed several people, and a couple of them have the experience and skill you're looking for. How do you decide which to hire? Here are a few other things to look for in evaluating your applicants.

Clothing — No one paints for eight hours and stays perfectly clean. Splotched, paint-splattered clothing comes with the territory. But paint-splattered clothing doesn't have to be tattered and dirty. Anyone who shows up for an interview in dirty clothes will show up that way on the job, too. And anyone working on your jobs represents your company. Sending a painter to a client's house in ragtag clothing makes a poor impression. This is

Finding and Keeping the Right People

Employment Application Form

Name _____ Home phone _____
Address _____ City/Zip _____
Education: High School _____ College _____ Trade School _____

List your last three employers:

Company name _____ Phone _____
Address _____
Supervisor's name_____ Employed from_____ to_____

Company name _____ Phone _____
Address _____
Supervisor's name_____ Employed from_____ to_____

Company name _____ Phone _____
Address _____
Supervisor's name_____ Employed from_____ to_____

Check the types of work you have experience in:

	How long		How long
___Exterior	_____	___Custom staining	_____
___Interior	_____	___Spraying lacquer	_____
___Custom	_____	___Spraying flat paint	_____
___New construction	_____	___Spraying enamel paint	_____
___Staining	_____	___Large industrial projects	_____

How you ever been a foreman or in charge of job production?_____

If so, what type of work? _____

How many total years of experience do you have? _____

Do you have a dependable car? _____

What tools do you have? _____
_____

Are you looking for full time work?_____ Part time work?_____

Are you willing to work part time? _____

Are you willing to work on Saturdays if necessary? _____ Evenings?_____

Are you willing to travel out of town to work? _____

What is the minimum hourly starting wage you would accept?_____

What was your hourly rate on your last job? _____

Person to contact in case of emergency or accident: _____

Employment application form
Figure 2-1

true in all types of painting, but it's especially important when doing custom residential work.

If you and your men show up on the job looking like bums, your client's going think, "Oh boy, these guys are slobs! That's probably the kind of work they do. I'd better watch them like a hawk." Don't let that happen to you. Make sure all of your employees know your policy: They should carry a change of "whites" (white T-shirt, shirt and white pants, overalls or jeans) in the truck. Encourage them to change whites occasionally during the job.

Painters aren't lawyers. No painter needs to be spotless. But people feel better about themselves and look more professional when neatly and appropriately dressed. That means a change of whites occasionally.

Neatness on the Job — Neatness is one of the standards I use to judge the professionalism of my painters. Your clients will use neatness as a standard by which your company is judged.

I can see two major benefits of neatness. First, neatness improves morale, productivity and promotes safety. Second, neatness gives clients the impression that your company is capable and professional. Let's take a closer look at both of these points.

Anyone working in a clean, well-organized area can accomplish more. Less time is wasted looking for tools and materials. Less time is wasted repairing soiled or spoiled work. There's less chance of an accident. Slovenly work habits can be downright dangerous if you're working in a home with children or pets. And most people feel more content and comfortable working in a clean, uncluttered environment.

Neatness means that tools are kept in good working order and are returned to an appropriate storage place when work is finished. It means that painting materials are clearly labeled and easy to get to. Buckets are kept clean for future use. There's nothing more frustrating than searching in vain for a clean bucket when you need one. Time spent searching is time wasted.

Good organization and neatness promote higher morale. That improves productivity and raises profits. By itself, that's enough reason to run a clean, neat operation. But the second reason makes the argument compelling: Neatness promotes your reputation as a highly professional, well-organized painter.

Picking up after yourself on a job always leaves a good impression. This is especially true when you're working in someone's home. Some clients don't know enough about painting to spot top-quality workmanship. But everyone knows a sloppy painter when he sees one. Leaving drop cloths, tools and paint scattered throughout the house gives your client the idea that you do sloppy work. Neatness on the job gives the impression of quality.

Several years ago my company conducted a survey of both our general contractor and homeowner clients. Among other questions, we asked, "What is the most important quality that a painting contractor should have?" The answers surprised me. I assumed that speed or quality or low price would be rated as number one. But of the clients who responded, 95% said neatness was most important. Your clients probably feel the same way.

Communication — One of the things I evaluate in a prospective employee is his ability to communicate. Notice that there's more to communication than just talking. Some nonstop talkers aren't communicating at all. Communication requires listening, understanding and responding. A good communicator needs both good listening skills and the ability to get a message across to someone. To communicate is to make a connection. A connection has to be a two-way street.

Your employees will usually be the first to notice something going wrong on a job. They'll be the first to hear a complaint when someone else notices a problem. Evaluating the problem, solving it, or telling someone about the problem takes communication skills. A painter who won't communicate or can't communicate is going to leave some problems unresolved and will make the rest worse.

Here's an example. One of your employees is working alone on a job. The client shows up on the job and starts complaining about a color match. At least three things could happen:

First, the painter could ignore the client and go on painting. He says to himself, "It's not my problem." This is obviously a mistake. It rates a flat F in my grade book. Ignoring the problem shows poor communication skills. I expect my painters to do better.

Second, he could refer the problem to someone. Just calling the office and passing on the complaint is a step in the right direction. That rates a C or maybe a C-plus if the painter showed courtesy and understanding in dealing with the client.

Third, the painter could use good communication skills to handle the problem right there. He could explain that paint changes color as it dries. The finished job will look more like the sample when fully dry. He could explain to the client that it's nearly impossible to match new paint with weathered paint, even when the formula is the same and the paint is from the same manufacturer. Further, he could point out that the contract for this job states that color samples are approximate and matches with existing paint are not guaranteed. There may be no way to produce an exact match. That's the type of response that rates an A for communication. In effect, the painter has done all that I could do. He's saved me an inspection trip to the job. And if he's satisfied the client, he gets an A-plus.

Good communication seems so simple. But so does painting — until something goes wrong. Then you've got a real mess on your hands. My advice? Stack the deck in your favor. Give hiring preference to painters who have good communication skills.

Hire the Qualified
It's a mistake to hire inexperienced people just because they'll work for lower wages. It's hard for new paint contractors to understand this. But most seasoned pros will agree that it's true. Nothing undermines crew productivity and quality quicker than an incompetent painter. Generally, you get what you pay for when hiring painters. There are bargains. But they're found at the top end of the pay scale, not at the bottom. Some exceptionally productive craftsmen earn more than the top wages they command. Very few underpaid novices are worth what they cost.

The key to making a profit in the paint contracting business is doing it right the first time and in the shortest time possible. With a novice, neither is likely. Here's an example. You sign a contract to paint the exterior of a home. The contract price is a little on the low side, but work's been slow lately. You know that there's still some profit in the job if the labor cost is kept to a minimum. To save money, you hire a couple of would-be painters at $6.00 per hour. You know that they don't have much experience, but you'll use them and keep your fingers crossed.

That's usually a mistake. Here's what happens nine times out of ten. The job starts out late and runs behind schedule from start to finish. Everything has to be done two or three times and still isn't right. There'll be a lot of touch-up at the end. You'll waste valuable production time showing your inexperienced painters what to do. The client recognizes that you're struggling and assumes that your company always works that way.

If you had used qualified painters, the job would have gone straight through with little touch-up. Your client would appreciate the speed and accuracy of your crew and recognize their professionalism.

Novice painters always take longer than an experienced crew. In general, the longer a job drags on, the more problems you can expect. Your crew should move onto the site with all the equipment and materials needed, proceed systematically and rapidly through prep, application and cleanup, and then pull off the job promptly. A brisk pace is the most effective way to minimize problems and maximize profits. I've never found novice painters who could maintain anything like a brisk pace.

Rewards and Incentives
Getting high productivity from each employee should be one of your primary objectives. A productive worker will give you a good day's work for a day's pay. But everyone on your staff could work better, faster and smarter with the right kind of incentive.

Incentives take many forms. Some cost nothing. Praise and thanks can be very effective. But I've found that money is the most universal motivator. Any tradesman who knows that there's a bonus for premium performance will usually work to earn that bonus.

Try this incentive on your next job. Tell your crew that finishing the job in four days instead of five will earn a $40 bonus for each painter and a $50 bonus for the crew leader. You may be surprised at how much productivity improves.

The bonus doesn't have to be a great deal of money. The reward is mostly in participating in the profits on a job well done. If the crew misses the four-day target but finishes before the end of the fifth day, let them know you appreciate the hard work. Give them a small bonus anyhow. Good production should be rewarded.

Here's another incentive you may not have tried. Let a crew "bid" on the job, quoting a total labor cost for the work. You furnish all tools, materials and equipment. You pay a set figure for their labor — no matter how long it takes. You're still the employer, responsible for job quality, handling the taxes and insurance, and collecting from the owner or general contractor. But the tradesmen are free to work at their own pace.

I wouldn't try this "bid" system with anyone but an experienced crew, men you know and trust to do good quality work. Neither would I use this system on anything but the easiest and simplest jobs, which can be done in a week or less. Anything more complicated will involve changes and negotiations that make a dispute more likely. And I wouldn't use this system on any job where there's the slightest doubt that full payment will be received immediately upon completion.

Please don't misunderstand what I'm saying. Under this "bid" system you're not subcontracting the labor. You're just setting the wage by negotiation before work begins. There's a big difference.

Taxes and Insurance

I've seen painting contractors try to "subcontract" labor. It doesn't work. Your state and the federal government make every employer responsible for insurance and taxes, no matter how the amount due is computed. Failing to withhold taxes and carry the workers' compensation insurance required by law subjects you to major penalties. Don't try to evade this responsibility.

Here are the taxes and insurance you have to pay on all employees, no matter how employees are paid:

State Unemployment Insurance: All states have unemployment insurance programs that provide benefits when a worker is laid off or disabled. The cost varies by state and with the employer's history of layoffs. But it's usually about 4% of gross payroll. You as an employer have to pay this to the appropriate state agency monthly or quarterly.

Federal Unemployment Insurance: The federal government also has an unemployment insurance program that provides extended benefits to all employees. You know this program by the initials FUTA. The cost is about seven-tenths of one percent of payroll (0.7%). You have to pay this tax quarterly.

Social Security and Medicare: This one really hurts. It's called FICA on employee pay stubs. Your share is about 7% of gross wages. Each employee also has to pay about 7% of gross wages. You kick in your 7% and withhold the employee's 7%. Deposits go in twice a month to a federal reserve bank.

Workers' Compensation Insurance: State law requires that you provide coverage for accidents which injure employees on the job. The cost of this insurance varies by state, by trade, by year and by your loss experience. For some trades in some states the cost of insurance is close to one-third of the base wage! For painters and paperhangers, the cost is usually between 7 and 10% of the base wage. The rate for supervisors and office workers is much lower, usually less than 1%. But the insurance companies won't divide an employee's time between two jobs. An employee who paints part-time and does office work part-time will be classified as a painter for workers' comp purposes. Premiums for workers' compensation insurance are paid to your insurance carrier monthly, quarterly or annually, depending on the policy you carry.

Liability Insurance: Every employer needs liability insurance to cover injury done to the public by employees. The cost of this insurance varies with the coverage needed and is based on gross payroll. The cost of adequate coverage will usually be about 2.5% of payroll. This is generally paid annually.

The total tax and insurance burden is more than 20% of payroll for nearly all painting contractors. And you as an employer pay it all. You also have to withhold state and federal income taxes from employee wages, but this comes out of the employee's pocket, not yours.

Is there any way to cut corners on these taxes? Not legally. Every employer has to pay in full and on time. Can you claim that your workers are independent contractors responsible for their own taxes and insurance? Not likely. If your state is like mine, you'll have a visit from an auditor every few years. The auditor has heard every excuse for not withholding taxes and has a good argument against each one. He's backed up by a legal department that's expert at bringing employers into compliance with the law, retroactively if necessary.

My advice? Pay your taxes. Figure your tax and insurance burden at 21 to 25% of labor cost on every job. If you're paying a painter $8 per hour, your cost is $10 per hour. Add 25% into every labor estimate, no matter how your tradesmen are paid.

When to Give Pay Raises

Every company with more than three or four employees needs a written policy statement on pay raises. Employees should know what they have to do to earn more money. Having a written policy avoids a lot of disputes. When you hire someone, explain exactly what's required to earn each pay raise. That gives new employees a goal to shoot for.

My recommendation is to grant raises only for increased:

1) Production (the volume of work completed)

2) Reliability (doing the job without assistance)

3) Responsibility (accepting more duties)

If you have an employee who's producing more, is more reliable and is accepting additional responsibility, he's worth more. Do what's necessary to keep that man on your payroll. This type of employee is adding to profits and contributing to company success. It's reasonable that he participate in the success of the company. Provide rewards that are in proportion to the employee's contribution.

Your policy on pay raises should explain how employees are evaluated. Make sure each employee has a copy of the policy statement. Here's an example:

Pacific Painting Job Evaluation Criteria

Pacific Painting employees are skilled professionals who provide high quality painting and coating services to the community. Employees are encouraged to accept additional responsibility, improve their reliability and find innovative ways to meet customer needs at lower cost and in less time. Pacific Painting grants pay raises to employees who can consistently demonstrate increased productivity, who show higher reliability and who accept additional responsibility.

Increased Productivity *means getting more work done in less time.*

High Reliability *means following company policy so that work is done correctly, on time and without assistance.*

Additional Responsibility *means accepting extra tasks and following them through to successful completion.*

No increases will be granted solely for longevity.

I think this last point is important. There's no advantage by itself to growing older in a job. An employee who doesn't increase productivity, show higher reliability or accept additional responsibility isn't worth any more than the day he walked in the door.

Suppose you tell Joe Smith that he's due for a dollar an hour raise after each six months of service. Joe has little incentive to improve his work. At the end of six months, will Joe be working any harder or contributing more to the company? Probably not. Why should he? You've robbed him of all incentive to produce more, improve his reliability or accept additional responsibility.

The Chain of Command

If you've ever served with the armed forces, you know the meaning of that term. It's the pecking order in any organization, the order of authority. In a painting company, the chain of command begins with the owner or owners. Below him is the field supervisor, who gives instructions to the job foremen. These foremen direct journeyman painters, apprentices and helpers. Let's take a closer look at each of these positions.

Owner— The buck stops here. The owner is ultimately responsible for everything that happens in the company. He answers only to himself, his partners or co-owners. In a small painting company, the owner may also wear the hat of field supervisor. In a two- or three-man company, the owner is also the foreman.

Field Supervisor— He oversees all production. The field supervisor schedules all jobs, keeps tabs on each, handles problems referred to him either by the owner or the foremen, and sets production targets for the foreman on each job. He reports to the owner. The field supervisor is usually nonproductive labor — he doesn't do any painting.

Smaller painting companies don't need a field supervisor until there are at least two crews working at the same time.

Foreman— He's normally in charge of only one job at a time. The foreman runs his crew, ensures that the crew meets production goals for each day, and makes sure that all necessary materials, equipment, and tools are available when needed. He handles production problems when possible and refers what he can't handle to the field supervisor or the owner. He should be able to do any type of work required on any job. Foremen are considered productive labor because they're painting when not directing other painters. The foreman reports to the field supervisor.

Journeyman Painter— He takes instructions from the foreman assigned to his job. He may be switched from one job to another as job requirements dictate. His job is pure production — painting!

Apprentice Painter— He's still learning to be a painter. He takes instructions from the foreman and the journeyman painters. His job is to learn the trade while he contributes to the productivity of the crew.

Helper— Every significant job includes hundreds of tasks that don't require skilled painters: moving supplies and equipment, protecting adjacent areas, moving drop cloths, carrying ladders, doing cleanup, going for coffee, sanding, running to the store for supplies. These tasks should be done by the lowest-paid man on the job. That's why it's a good idea to include a helper on any crew that has more than two or three tradesmen.

Helpers are the bottom link in the chain of command, but they can save you hundreds of dollars in painter time on a large job. Use helpers whenever possible. They keep your painters painting.

The chain of command provides a route for the flow of instructions and information going up and down the chain. Bypassing a link in the chain is usually a mistake. For instance, if there's a problem on a job, a journeyman shouldn't start by looking for the owner. The journeyman should follow the chain of command, reporting the problem to the foreman. If the foreman can't handle the problem, he takes it to the next person in the chain of command, usually the field supervisor.

From there the field supervisor may take the problem to the owner.

Instructions flowing down the chain should also hit every link. The owner who makes it routine practice to bypass supervisors and foremen is undermining the authority of those passed over. In an emergency, anything goes, of course. But routine instructions should stick to adjacent links in the chain.

In spite of the value of having a chain of command, there are times when ignoring it is better than using it. I allow employees to skip the chain of command when it comes to personal problems or disputes that only the boss can settle.

My door is always open to anyone in the company who has a problem that's affecting his work. For example, I don't insist that a journeyman or helper discuss a serious personal problem with the foreman or supervisor before bringing it to my attention. But if the problem involves a supervisor or foreman, I'll call that man into the conference.

Accidents

Every experienced painter has probably met an accident-prone painter. Things just keep going wrong around him. He may have the best of intentions and try very hard. But for some reason every time he sets foot on the job, a bucket of paint gets dropped or a roller pole breaks a window.

Anyone can have an accident. But the best workers have less than their share. Others sometimes seem to have far more than their share. If there's someone on your payroll who's accident-prone, you may be better off without him — even if he's an excellent painter.

Every employer wants to show understanding and compassion toward employees. But it's foolish to keep a high-risk painter on the payroll. A little accident with a spray gun can cause a serious injury or cost you the price of a car paint job. Figure 2-2 shows the kind of event your business can well do without. All joking aside, this happens, and can cost you hundreds or even thousands of dollars — and most likely the goodwill of your client.

When to Fire

The last two topics that I have to cover in this chapter aren't my favorites. Neither will they be yours. But they're important enough to deserve your attention. The first is terminating employees.

You fire employees for one of two reasons: First,

Finding and Keeping the Right People

**Accidents are costly
Figure 2-2**

there isn't enough work to keep everyone busy. In that case the discharged employee won't be replaced. Second, you want to get rid of a particular employee. In that case the employee *will* be replaced.

Laying off loyal, trusted employees for lack of work isn't easy. But at least you can offer sympathy and maybe a job later when work picks up. Firing an employee for misconduct or poor work is harder. When and how to fire under these circumstances is a matter of intuition as much as anything. But I can offer some guidance.

Looking back over the employees I've fired in the last ten years, my only regret is that I didn't fire some of them sooner. There are square pegs and there are round holes. Trying to force a square peg into a round hole is a mistake. Some painting companies and some painters just aren't meant for each other. When you discover that situation, the sooner you part company, the better — for both your company and the employee involved. The Yellow Pages are full of other painting companies. Your discharged employee may fit in perfectly in one of them.

When I reach a decision to fire someone, it's always because that person has not followed the company's policy statements. These statements establish what's expected of all employees. They're the rules every employee has to play by. Anyone who isn't following the rules should expect to be discharged.

If someone has trouble complying, give him a private verbal warning. Make a note of that warning and the violation that made the warning necessary. A second violation might be worth a letter of reprimand. If there's a third violation, the employee should know what's coming. In effect, he's fired himself by refusing to comply with company policy.

When it comes time to fire someone, do it as directly and as promptly as possible. Don't be apologetic or indicate that there's any chance you'll change your mind. You're not doing anyone a favor by making excuses. When it's done correctly, an occasional firing clears the air. It makes the employees remaining on the payroll appreciate their jobs more. And it gives everyone in the company a better understanding of your role and their role.

One last point about firing. Do it fast. Don't let it drag out over several days. Once you've made your decision, take the individual aside as soon as possible. Pay him and send him out the door.

Employee Dishonesty

The last subject in this chapter is employee dishonesty. In my book, an employee that steals from the company or cheats the company is off the payroll, immediately and permanently. I imagine that you agree.

The more difficult problem is dishonesty that can't be proved. I have no simple answer here, but suggest that a pattern of disappearances may point so strongly to some individual that the company is better off without him. This is true no matter how good a painter he may be. When you're painting the inside of someone's home, the last thing you want is a painter going through the homeowner's desk looking for valuables. Take my advice. If you identify a dishonest person, get rid of him.

Chapter 3

Putting on a Good Face

There are two types of paint contractors:

The first type does commodity grade work — applying coatings at the lowest cost, for the lowest price. His competition is every cutthroat painter in town. He takes jobs with little or no profit just to keep busy. And he often ends up making less than if he were working for wages on some other contractor's payroll. This painter knows he's underpricing his work. But he hopes to make it up in volume. In truth, he seldom does. His preoccupation with high volume and low costs make it impossible to zero in on what customers really want, and what they're willing to pay for. He goes on, year after year, earning enough to get by but never earning a reputation for craftsmanship. He'll probably never accumulate enough capital to break into the ranks of the big-time painting companies.

The painter I've just described has no interest in good client relations. If you're determined to be this type of painter, skip over this chapter. It has nothing that will interest you. But if you want more from the painting business than just a job, keep reading.

The other type of painter sees his job from a different perspective. He doesn't just apply coatings and furnish labor. His job is service. His product is satisfied customers. It just happens that he satisfies customers by applying paint and coatings. Though his prices are competitive in the class of work he handles, they're not the lowest. But no one can beat him in meeting customer needs. He's tops in selecting color, suggesting the right surface preparation for the job, recommending top quality coatings, and selling his service as a quality painter. In short, he specializes in customer service as well as painting. Maintaining good relations with his clients is as important as anything that happens in his company.

If you're thinking that you would rather paint than try to satisfy customers, maybe you should. But I guarantee that the rewards are better for any painter who can learn the points that I'm going to cover in this chapter. If you're willing to learn, keep reading. I'll make it worth your time.

Keep in mind that I'm not claiming that the customer is always right. Far from it. The customer may be dead wrong. You may be right. If so, and if you have the facts to back you up, be polite but stand your ground. There's nothing unprofessional about standing firm when a client is trying to abuse your good nature.

Trust is Basic

Satisfying customers means more than just living up to the requirements in a contract. It would take 50 pages of fine print to cover every potential point of disagreement on even a simple painting job. I've never seen such a contract and you won't either. Until that contract becomes standard, every job will be based on trust — trust that you will do a thousand things. Things like paint every surface that should be painted, keep paint off what shouldn't be painted, and clean up when you're finished. Trust has to cover a lot: quality of work, schedules to keep, services to perform, products to deliver.

No contract is complete enough to make trust unnecessary. That's why clients have to trust your integrity. And that's worth plenty to many, and probably most, clients. A customer who's confident that you'll do a professional job and live up to his expectations is willing to pay for that peace of mind. Clients trust contractors with a reputation for good customer service. And that's exactly the reputation you want to build.

Working with a client is like any other relationship. It can be pleasant and rewarding, or filled with distrust, misery and frustration. Which type of relationship it is depends largely on you.

Of course, there are some customers who refuse to get along with anybody. When you run into one of these characters, it's best to just get the job done as quickly as possible and get out of there. Clients like this, however, are the exception, not the rule. The vast majority of them are decent people, and you can establish a good working relationship. There will be occasional upsets and problems. But if you're patient, a good listener, and willing to handle problems as they occur, you'll be able to keep things running smoothly. That's the key: a willingness to work with your customers.

Make good customer relations and a quality product the foundation of your reputation. It's an unbeatable combination for success.

Anyone who hires your company has two objectives: a satisfactory job and pleasant, problem-free relations with you. Everyone has enough problems already. No one wants more. This is worth keeping in mind when you start a job.

Any time you take on a new client, make it your policy to keep that relationship positive. This is the basis for building an excellent reputation and abundant referrals.

Most painting contractors deal with two types of people: homeowners and professionals in the construction industry. Professionals include contractors, architects, and interior designers. They know the business and its peculiarities. Each type of client comes with a particular set of problems. Homeowners generally have a wider range of problems. But the recommendations in this chapter apply to both groups.

Show Them You're a Professional

People respect professionalism. Look like a professional, act like a professional, and most clients will deal with you as a professional. This is a crucial first step in customer relations. When you gain a client's confidence, he'll respect your opinion and judgement.

The public's image
Figure 3-1

Many people expect painters to look like the guy in Figure 3-1. They judge you first by your appearance. The professional in Figure 3-2 is going to make a much better impression. A neat and clean appearance promotes trust. This is to your advantage when dealing with any customer.

Here are some other characteristics which show your clients that they're dealing with a professional.

Good Manners are Good Business

Having good manners means being respectful and courteous with your customers. It means being patient, listening to what they have to say and answering their questions. It means conducting yourself properly while in their home, being polite and being careful in your language and remarks.

The reason to use good manners with your customer is simple — it's good for business. Practice good manners. You'll get along better with your customers.

Never Allow Booze on the Job

Any alcoholic beverage on the job plants a seed of mistrust. I can see four good reasons to ban alcohol from every job you have.

First, even one beer reduces both the quality and quantity of work done.

Second, alcohol promotes rowdy behavior, increases the risk of an accident and makes damage to your client's property more likely.

Third, it looks bad. If a problem comes up and there was alcohol on the job, the contractor who didn't ban all booze will always take the blame. If you should ever have to take a customer to court to collect, and the customer can say the painters just sat around and got drunk (even if they just had one beer), you may as well forget about getting paid for *that* job.

The fourth reason is elementary. You're there to work, not party. It doesn't make good sense to pay someone to drink when he should be working.

Drinking on the job is asking for trouble. Don't allow it. If your painters are accustomed to opening a beer at quitting time, insist that they leave the job before popping the first can.

Honesty Pays

Honesty is always the best policy. No business can succeed for long if it gets a reputation for dishonesty. Don't be tempted to make a few quick bucks on one job. In the long run, you'll be the loser. Nine out of ten customers can spot dishonesty. Even if they can't prove it, they'll scrutinize everything you do and avoid you in the future.

People truly appreciate and respect honesty. It's expected of every real "pro" in the business. The payoff is in referrals, a favorable reputation, and repeat business.

Troubleshooting is Your Job

The ability to handle problems is valuable in every business. It's especially valuable in the painting business. As a painting contractor, you'll have days when you begin to feel like a lightning rod attracting every stray thunderbolt in the area. Wading through the problems and smoothing ruffled feathers may take much of your time. But there's no alternative. Ignoring a problem, dodging responsibility or imposing an unfair solution will only preserve the problem until it becomes an even larger irritation later on.

Problems are inevitable. Your employees will have differences with you, with each other, with clients, and with their own families. You have your own dilemmas. And so does your client. There's no way to avoid all problems. Some jobs are plagued with upsets. Others seem to be relatively free of trouble. But you'll have at least a couple of problems to resolve or head off on most any job.

The easiest way to avoid problems is to anticipate them. Once you spot a potential problem, do whatever's necessary to keep it from becoming an actual problem. Of course, that isn't always possible. But your policy statements will help avoid

A professional image
Figure 3-2

many problems. Employees know what's expected of them and what they can expect in return. That eliminates many surprises with employees. And it's surprises that cause most of the problems you'll have with employees.

Your contract should eliminate most surprises with clients. If the contract is reasonably complete and if you decide in the beginning to leave the customer satisfied, there won't be many disputes. Staying on schedule and doing a professional job creates a lot of good will.

Here are some ways I've found to resolve most upsets, problems and misunderstandings.

First, always listen to and acknowledge what your customer has to say. Try to understand his position and point of view. Communication is, after all, a two-way street. Sometimes the simple act of listening respectfully will solve the problem all by itself. Of course, some people are accustomed to being ignored. They may be so delighted that you're actually listening that they'll never stop talking.

Next, explain your position. Some homeowners assume that they're being cheated by every contractor. They'll probably become suspicious and agitated if they don't understand what you're doing. So be patient and explain. Five or ten minutes of your time spent explaining something to your customer may save an hour of arguing when it's time to collect.

The benefits of good communication are amazing. If you truly want to get a problem settled, just keep on communicating until a solution is found. Of course, this isn't always easy. It takes nerve to confront an irate customer face-to-face. It's also unpleasant. But resolving problems is part of your job. It's something you'll have to learn to do if you want to succeed.

Deliver What You Promise
Success is built on delivering what's promised. You've got a contract to deliver a service and a product. Be professional and deliver it. If you have to start a job by May 5, don't wait until the 4th to start planning. Sketch out in your mind the start date, material delivery date, and the finish date. Plan to meet each in succession.

If you run into delays, get in touch with your customer. Keep him informed of what's going on. He'll appreciate it. Wondering what's happening is worse than knowing how long the delay will be.

In reality, all of your business is based on trust and good will. In most cases, the cost of resolving conflicts in a courtroom will be more than the amount in dispute. Protect yourself by keeping your promises and letting the client know immediately when you can't.

Be Firm in Your Decisions
Your company is a professional organization. It has written policies. It follows established schedules. It has standards of quality. It observes certain obligations and responsibilities to both staff and clients. These standards and policies ensure that everyone is treated fairly and in a professional manner. You abide by these standards to protect both clients and employees. Anyone who asks you to ignore some established standard is asking you to violate an important principle for his own benefit. You're under no obligation to comply with a request like that.

I've had customers ask for special consideration many times. Starting a new job before another is finished may work out fine for one customer. But it disrupts your schedule, angers another customer, and adds to production costs. My advice? Easy. Be as courteous as possible, but also be firm. Point out the advantages of dealing honestly with all customers and staying on the planned schedule. Your client may be unhappy. But he will also be impressed with your consideration for other clients.

You also need to be firm in the matter of receiving payment. On larger jobs, try to avoid potential problems by setting up a schedule of progress payments. Ask for an initial payment as soon as materials arrive on the job site. That helps cover the cost of materials. Get a partial payment at the end of each week until the job is completed. If your client falls behind the payment schedule, threaten to pull your crew off the job until payment is made. If the payment schedule is in writing, you have every legal right to do this. You also have the responsibility to yourself, your company and your employees to collect the money owed you.

There are always exceptions, but it's reasonable to be firm and direct when a client misses even one payment. Ignoring that oversight is an invitation for him to miss the next payment. Also, note that a missed payment raises a danger signal that the client may be having financial trouble. Whatever his intentions, he may never be able to pay even if the job is completed on time.

Stick to your guns. You establish company policies, operating procedures, schedules and product quality standards. If your clients make these decisions, you've put the company into their hands. Only an amateur would make that mistake.

Make it clear that you run the company by rules that are designed to protect both clients and employees. Set standards and then stick to them. The public will respect this. Anyone who hires a craftsman wants him to take charge and expects him to act like a professional.

Stick to the Schedule
Setting a schedule, explaining it to your client, and then sticking to that schedule is a simple matter of good manners. It's also the rule of construction etiquette most commonly violated. How often have you heard someone complaining that the plumber didn't come, the painters haven't shown up, or the carpenter is overdue?

Being a no-show is rude. Leaving a customer in the dark about the no-show is worse yet. Telephone calls are very cheap. When you can't meet a schedule, it's no crime. Just give your customer as much advance notice as possible. That's the professional way to handle it.

Missing schedules is a fact of life in the construction business. Too many things can go wrong — weather, illness, delayed delivery, underestimating the work time required. Getting off schedule occasionally is unavoidable. But if you're always running several days late, something is wrong. A schedule isn't a schedule if it's only wishful thinking. Don't burden clients with your dreams. Give them realistic schedules you can meet.

If you can't start the job on the start date, contact the customer and let him know as early as possible. If your crew needs a day off for personal reasons, call the customer as soon as you know about it. Your customers have schedules and plans too. As a matter of courtesy (and good customer relations), advise clients of any absence or delay in the completion of their job.

The Extra Touch
It's good policy to do just a little bit more than the contract or your agreement with the client requires. I call this the "baker's dozen" approach. The bakery where I buy doughnuts always gives me 13 doughnuts when I order a dozen. Maybe that's why I keep going back for more. No matter how many times I've been in that shop, I still leave feeling like I've just won a free doughnut. Maybe you know the feeling.

Once the counter girl at the doughnut shop made a mistake. I actually got only 12 doughnuts, not the usual 13! Did I complain? Not a bit! I couldn't. After all, I only paid for a dozen.

It's the same way with your paint jobs. Do slightly more than you have to. Paint an extra door at no charge. Rearrange the furniture after the room is painted. Clean out a few paint cans and leave them for your client. Touch up some spots on a baseboard that you didn't even have to paint.

The extra touch
Figure 3-3

Leave a small jar of the paint you used so your client can do touch-up later. Find something "extra" you can do that costs little but adds to the value of what you've done. And always do it with a smile, like the "movers" in Figure 3-3. It's at no charge, of course, so your client feels like he got something for free. It also makes quibbling over some minor discrepancy much less likely.

Of course, there's a limit to how much extra you can do on a job. If a client asks me to paint some trim, or a soffit that wasn't included in my bid, I'm happy to do the work — but I charge at my usual hourly rate. The "baker's dozen" approach works only when you do the work voluntarily. A client asking for extra work will have to pay the extra charge. Some people would have you do the whole job at no charge if you're silly enough to do it.

Extra touches are good for business. But don't lose sight of the objective. You're in the painting business to make a profit and earn a good living, not as a charity.

Beware of "Friends"
When you're making an estimate, you'll probably have some potential client request a cut rate because he has lots of friends that he'll refer you to. As soon as you hear this, add 10% to the bid. Giving this customer a cut rate is cutting your own throat. If he ever refers any business your way, it's because he's passed the word that you're a lowball operator. Preserve your reputation as a quality painter who works at fair but firm prices. Stay away from "deals" with fast-talking chiselers.

Rely on Written Agreements
Thousands of contractors and craftsmen take on jobs without written contracts every day — and every day there are thousands of problems.

My advice is to get a written agreement on every job over $5. If you have nothing in writing, then everything is left to memory and your client's sense of fair play. Depending on memory usually leaves you with "I thought you said" and "That isn't what I meant." When that happens, the client has the leverage because he still has the money and you've done the work. When there's nothing in writing, the customer is usually going to get his way in a dispute.

Have a standard contract with lots of fine print on the back and a place to fill in the customer name, address, price and description of the work on the front. You sign it and your client signs it. Each of you keeps one copy. Many stationery stores carry contract forms. If you don't have a contract form and can't locate one you like, consider using the contract in Figure 3-4.

Avoid surprises and you avoid disputes. A written contract that lists exactly what's to be done will avoid most surprises — no matter how large or small the job is.

Good contracts make satisfied customers. Putting the job description in writing forces the owner to describe exactly what he wants done. Doing more will cost more, anything less and the work's not done. Avoid verbal agreements. They're like putting the future of your business in a stranger's hands.

And beware of the customer that doesn't want to sign a written agreement. Anyone who won't put his promises in writing either doesn't understand the law or doesn't want to be held to an agreement. If he won't sign, don't work for him. His motives are too questionable.

On a large job, have the customer fill out a credit application when he signs the contract. You're granting credit to a client when you do the work before getting paid. Be sure the client is creditworthy before work begins. Figure 3-5 is a good credit application.

Keep a Job Log
Every professional painting contractor should keep a job log. I've used one for years and wish I could claim credit for inventing the idea. But I can't. It probably originated with job superintendents on large construction projects. The book I use measures 3½" by 6" so it fits in my shirt pocket. It's spiral bound with a plastic cover and has a page for each day. You'll find a selection of small notebooks like this at any good stationery store.

In my book I record the work done each day on every job. Only a few words are needed to summarize progress on each job. But I add notes about anything else that happens which may be important later. Any time I visit a job, I make a note of the time I was there and what I found. On an average day, I might spend five or ten minutes making notes in my job log. The time invested is very little compared to the value of this record.

My job log is the official record of what happens every day in my company. It has settled more disputes than I can remember. Here's why. First, I have the only written record of what happened. Others can only guess about who did what that day. I have notes taken from my personal observations, in my own handwriting, and written at the time the observations were made.

Second, my notes are admissible as evidence in court. The legal system puts real trust in systematic notetakers who record personal observations regularly.

Here are my entries for Monday, May 5:

Nelson crew started King job, set up & prepped ext. all day.

Lopez crew prepped & painted ext. stucco on Smith job – finished at 4 p.m.

At 10:15 a.m. Williams crew stopped ext. work on Johnson job at Mrs. Johnson's request & painted int. L.R., D.R. and kitchen to 4 p.m.

If Mr. Johnson grumbles later on about the exterior not being finished in time for his lawn party, the logbook entry will remind everyone that Mrs. Johnson had you move the crew off the exterior to begin the interior work on May 5.

Write an Operations Statement
I've found that an Operations Statement is a convenient way to inform clients of important facts about my company. The Operations Statement is a brief written statement of how the company operates. It's a problem-solving tool designed to help you and clients resolve problems before they become serious.

Many clients think that only the contractor-owner can solve any problem they may have. That may be true in many small painting companies. But if you have a partner and two or three crews, you can't possibly handle every minor problem yourself. An Operations Statement encourages clients to channel questions directly to whoever can solve the problem. You get involved only if that person can't provide a remedy.

Figure 3-6 is a sample Operations Statement. Give your client a copy when the contract is signed. A good Operations Statement will solve at least half of the common client problems that take up so much of a contractor's time.

The Contractors' Image
Not all contractors are good guys. It's the unethical or incompetent contractor who makes life difficult for you and me. There are just enough bad guys in the construction business to make the public naturally suspicious toward all contractors.

This is particularly true in remodeling work. By the time the painters show up, the homeowner may have endured months of upset, unsatisfactory work and delay by other contractors and craftsmen. The homeowner probably hates you before you walk through the door. The best way to disarm a hostile client is with your professionalism and attention to detail.

Listen to the problems a client has had. Agree with him when that seems appropriate. But don't run down your fellow contractor. There's no advantage in that. Point out that you're not the carpenter or plumber. Explain that you run a professional operation and place high value on customer satisfaction. Then follow through with the kind of job that anyone would agree is first rate and professional.

Proposal and Contract

Date _____ 19 _____

To _____

Dear Sir:

We propose to furnish all materials and perform all labor necessary to complete the following: _____
_____
_____
_____
_____
_____
_____

Job Location: _____

All of the above work to be completed in a substantial and workmanlike manner according to the terms and conditions on the back of this form and the detailed job estimate for the sum of:
_____Dollars ($ _____)

Payments to be made as follows: _____
_____
_____
_____

the entire amount of the contract to be paid within _____ days after completion. The price quoted is for acceptance within 10 days. Any delay in acceptance will require a verification of prevailing labor and material costs.

By _____

Company Name _____

Address _____

State License No. _____

"YOU, THE BUYER, MAY CANCEL THIS TRANSACTION AT ANY TIME PRIOR TO MIDNIGHT OF THE THIRD BUSINESS DAY AFTER THE DATE OF THIS TRANSACTION. SEE THE ATTACHED NOTICE OF CANCELLATION FORM FOR AN EXPLANATION OF THIS RIGHT."

You are hereby authorized to furnish all materials and labor required to complete the work according to the terms and conditions on the back of this proposal, for which we agree to pay the amounts itemized above.

Owner _____

Owner _____ Date _____

Sample contract
Figure 3-4

Terms and Conditions

1. Contractor shall begin work under this agreement and continue the work hereunder to completion within a reasonable time, subject to such delays as are permissible under this contract. Contractor shall obtain a valid building permit from the appropriate Public Authority if such building permit is required. Any fee or charge which must be paid to the Public Authority in connection with the work will be paid by Owner unless provided otherwise under this contract.

2. Contractor shall pay all valid bills and charges for material and labor arising out of the construction of the structure and will hold Owner of the property free and harmless against all liens and claims of lien for labor and material filed against the property.

3. No payment under this contract shall be construed as an acceptance of any work done up to the time of such payment, except as to such items as are plainly evident to anyone not experienced in construction work.

4. Unless otherwise specified, the contract price is based upon Owner's representation that there are no conditions preventing Contractor from proceeding with usual installation procedures for the materials required under this contract. Further, Owner represents that he will relocate furniture, clothing, draperies, personal effects, all personal property, plants, trees, and bushes prior to the beginning of work so that Contractor has free access to portions of the building where work is to be done. In the event that Owner fails to relocate any items as provided hereunder, Contractor may relocate any of Owner's property as may be required and is not responsible for damage thereto which may result during prosecution of the work.

5. Owner agrees to pay Contractor his normal selling price for all additions, alterations or deviations. No additional work shall be done without the prior written authorization of Owner. Any such authorization shall be on a change-order form, approved by both parties, which shall become a part of this Contract. Where such additional work is added to this Contract, it is agreed that all terms and conditions of this Contract shall apply equally to such additional work.

6. The Contractor shall not be responsible for any damage occasioned by the Owner or Owner's agent, rain, windstorm, Acts of God, or other causes beyond the control of Contractor, unless otherwise herein provided or unless he is obligated by the terms hereof to provide insurance against such hazards. Contractor shall not be liable for damages or defects resulting from work done by Subcontractors. In the event Owner authorizes access through adjacent properties for Contractor's use during construction, Owner is required to obtain permission from the owner(s) of the adjacent properties for such access. Owner agrees to be responsible and to hold Contractor harmless and accept any risks resulting from access through adjacent properties.

7. The time during which the Contractor is delayed in his work by (a) the acts of Owner or his agents or employees or those claiming under agreement with or grant from Owner, or by (b) the Acts of God which Contractor could not have reasonably foreseen and provided against, or by (c) stormy or inclement weather which necessarily delays the work, or by (d) any strikes, boycotts or like obstructive actions by employees or labor organizations and which are beyond the control of Contractor and which he cannot reasonably overcome, or by (e) extra work requested by the Owner, or by (f) failure of Owner to promptly pay for any extra work as authorized, shall be added to the time for completion by a fair and reasonable allowance.

8. Contractor shall at his own expense carry all workers' compensation insurance and public liability insurance necessary for the full protection of Contractor and Owner during the progress of the work. Certificates of such insurance shall be filed with Owner and with said Lien Holder if Owner and Lien Holder so require. Owner agrees to procure at his own expense, prior to the commencement of any work, fire insurance with Course of Construction, all Physical Loss and Vandalism and Malicious Mischief clauses attached in a sum equal to the total cost of the improvements.

9. Where colors, textures, shades or hues are to be matched, Contractor shall make every reasonable effort to do so using standard materials, but does not guarantee a perfect match. At Owner's written request, Contractor will provide a sample of any color, texture, shade or hue to be used under this contract for approval or disapproval by Owner. If Owner does no so request, Contractor is authorized to apply manufacturer's standard colors, textures, shades and hues as identified in this contract and is not responsible for any discrepancy between the manufacturer's sample and the material as applied.

10. Contractor makes no warranty, express or implied (including warranty of fitness for purpose and merchantability). Any warranty or limited warranty shall be as provided by the manufacturer of the products and materials used in construction.

11. Any controversy or claim arising out of or relating to this contract, shall be settled by arbitration in accordance with the Rules of the American Arbitration Association, and judgment in any Court having jurisdiction.

12. Should either party hereto bring suit in court to enforce the terms of this agreement, any judgment awarded shall include court costs and reasonable attorney's fees to the successful party plus interest at the legal rate.

13. Owner grants to Contractor and Contractor's employees and Subcontractors the right to enter the premises during daylight hours from Monday through Friday between 8 A.M. and 5 P.M.

14. The Owner is solely responsible for providing Contractor prior to the commencing of construction with such water, electricity and refuse removal service at the job site as may be required by Contractor to carry out this contract. Owner shall provide a toilet during the course of construction when required by law. Contractor shall leave living areas "broom clean" at the completion of work. Contractor shall make every reasonable effort to reduce overspray and paint splatter but is not responsible for damage to Owner's personal property that results from overspray or paint splatter.

15. The Contractor shall not be responsible for damage to existing walks, curbs, driveways, cesspools, septic tanks, sewer lines, water or gas lines, arches, shrubs, lawn, trees, clothesline, personal property, telephone and electric lines, by the Contractor, Subcontractor, or supplier incurred in the performance of work.

16. Work shall be completed and Contractor shall be entitled to prompt payment in full when the work described in this contract has been performed. Contractor is not obligated to do any work or perform any service except as expressly provided in this agreement. If, after Contractor has declared the work completed, Owner claims that work still remains to be done, Owner agrees to make prompt payment of the full contract amount less only an amount needed to hire a competent tradesman and purchase the materials needed to complete the work claimed yet to be done by Owner. Upon completion of any corrective work claimed by Contractor, Contractor shall be entitled to payment of the full contract amount.

17. Contractor has the right to subcontract any part, or all, of the work herein agreed to be performed.

18. Owner agrees to install and connect at Owner's cost, such utilities and make such improvements in addition to work covered by this contract as may be required by Public Authority prior to completion of work of Contractor.

19. Contractor makes no guarantee or promise concerning durability of materials or reduction in fuel bills as a result of any work performed. Heating and cooling costs are a function of utility rates, lifestyle, activities of the occupants, temperature at which thermostats are set, hot water usage, ventilation and many other factors over which Contractor has no control.

20. Contractor shall have no liability for correcting any existing defect which is recognized during the course of work.

21. Owner hereby grants to Contractor the right to display signs and advertise at the building site until all work is completed and payment in full has been made.

22. Contractor shall have the right to stop work and keep the job idle if payments are not made to him when due. If any payments are not made to Contractor when due, Owner shall pay to Contractor an additional charge of 10% of the amount of such payment.

23. Within ten days after execution of this Contract, Contractor shall have the right to cancel this Contract should he determine that there is any uncertainty that all payments due under this Contract will be made when due or that an error has been made in computing the cost of completing the work.

24. This agreement constitutes the entire contract and the parties are not bound by oral expression or representation by any party or agent or either party.

25. The price quoted for completion of the structure is subject to change to the extent of any difference in the cost of labor and materials as of this date and the actual cost to Contractor at the time materials are purchased and work is done.

26. The Contractor is not responsible for labor or materials furnished by Owner or anyone working under the direction of the Owner and any loss or additional work that results therefrom shall be the responsibility of the Owner.

27. No action arising from or related to the contract, or the performance thereof, shall be commenced by either party against the other more than two years after the completion or cessation of work under this Contract. This limitation applies to all actions of any character, whether at law or in equity, and whether sounding in contract, tort, or otherwise.

28. Contractor agrees to complete the work in a substantial and workmanlike manner but is not responsible for failures or defects that result from work done by others prior, at the time of or subsequent to work done under this agreement, failure to keep gutters, downspouts and valleys reasonably clear of leaves or obstructions, failure of the Owner to authorize Contractor to undertake needed repairs or replacement of water-damaged, blistered, peeling or otherwise deteriorating surfaces. Contractor is not liable for any act of negligence or misuse by the Owner or any other party.

Sample contract
Figure 3-4 (continued)

Notice To Customer Required By Federal Law

You have entered into a transaction on _____ which may result in a lien, mortgage, or other security interest on your home. You have a legal right under federal law to cancel this transaction, if you desire to do so, without any penalty or obligation within three business days from the above date or any later date on which all material disclosures required under the Truth in Lending Act have been given to you. If you so cancel the transaction, any lien, mortgage, or other security interest on your home arising from this transaction is automatically void. You are also entitled to receive a refund of any down payment or other consideration if you cancel. If you decide to cancel this transaction, you may do so by notifying

_____
(Name of Creditor)

at _____
(Address of Creditor's Place of Business)

by mail or telegram sent not later than midnight of _____ . You may also use any other form of
                                                        (Date)

written notice identifying the transaction if it is delivered to the above address not later than that time.

This notice may be used for the purpose by dating and signing below.

I hereby cancel this transaction.

_____   _____
        (Date)   (Customer's Signature)

Effect of rescission. When a customer exercises his right to rescind under paragraph (a) of this section, he is not liable for any finance or other charge, and any security interest becomes void upon such a rescission. Within 10 days after receipt of a notice of rescission, the creditor shall return to the customer any money or property given as earnest money, downpayment, or otherwise, and shall take any action necessary or appropriate to reflect the termination of any security interest created under the transaction. If the creditor has delivered any property to the customer, the customer may retain possession of it. Upon the performance of the creditor's obligations under this section, the customer shall tender the property to the creditor, except that if return of the property in kind would be impracticable or inequitable, the customer shall tender its reasonable value. Tender shall be made at the location of the property or at the residence of the customer, at the option of the customer. If the creditor does not take possession of the property within 10 days after tender by the customer, ownership of the property vests in the customer without obligation on his part to pay for it.

Sample contract
Figure 3-4 (continued)

Paint Contractor's Manual

CREDIT APPLICATION

Name _____ FIRST _____ MIDDLE _____ LAST _____ Date of Birth _____

First Name of Spouse _____ Social Security Number _____ - _____ - _____

Home Address _____ City _____ State _____ Zip _____

Home Phone (_____) _____ Years at Present Address _____ Own Home ☐ Rent ☐ _____

Married ☐ Single ☐ Divorced ☐ Widow(er) ☐ Number of Dependents _____

Previous Home Address _____ How Long? _____

Firm Name or Employer's Name _____ Years There _____

Address _____ City _____ State _____ Zip _____

Business Phone (_____) _____ Position _____ Nature of Business _____

Previous Employer _____ Years There _____

Address _____ City _____ State _____ Zip _____

College/University (if recent graduate) _____ Year Graduated _____

Your Present Annual Salary _____ List Source & Amount of income other than salary _____

Personal References: Name _____

Address _____ City _____ State _____ Zip _____

Name _____

Address _____ City _____ State _____ Zip _____

Credit References: 1. Name _____ No _____ Open ☐ Closed ☐

                        Address _____

                        2. Name _____ No _____ Open ☐ Closed ☐

                        Address _____

Bank 1. Name _____ Branch _____

        _____ Type of Account _____ No _____

Bank 2. Name _____ Branch _____

        _____ Type of Account _____ No _____

Finance Company Name _____ Address _____

Street _____ City _____ State _____ Zip _____

Nearest Relative or Friend Not Living with you _____

Address _____ Relationship _____

I hereby certify that the information in this credit application is correct. I hereby authorize you or your agent to investigate the data furnished by me.

        X _____
          SIGNATURE OF APPLICANT (INK)   DATE

Sample credit application
Figure 3-5

Doe Painting Company

Contractors: Joe Doe & Larry Doe.

Joe and Larry are partners and owners of the company. Either Joe or Larry will visit every job at least once a day. The time they spend there and the time of the visit will vary. But they can always be contacted by calling (508) 638-4700. After normal business hours, an answering machine will accept your message. Either Joe or Larry will return the call by 9 A.M. the next business day.

Estimator: Joe Doe

Production coordinator: Larry Doe

Office manager: Penny Driscol

For questions about price, additional work, or contracts, contact Joe Doe.

For questions about scheduling, production or quality of work, first contact the foreman on your job. If the foreman can't answer them, contact Larry Doe.

Payments are picked up by Joe Doe. For information about the amount you owe or payment dates, contact Penny Driscol at (508) 638-4700.

The basic steps we follow when painting are:

Surface preparation — patching, sanding, puttying, and priming where needed.

Finish coatings — applied as specified in the manufacturer's recommendations.

The foreman will explain the procedure we plan to use on your job.

We use only top-quality paints, coatings and finishes. But we cannot be responsible for *exact* matching of existing colors or textures. Neither can we guarantee that the paint samples you selected will *exactly* match the finished coating. All paints change color as they age. There is no way to predict exactly how age, sunlight and weather will affect any surface coating.

We will cover or otherwise protect adjacent areas, materials, plants and trees. We suggest that you remove all objects of value from the painting area. We will protect bushes and trees by covering, if possible, and tie branches back and away from the surface to be painted. This gives us room to erect ladders and scaffolds, and prevents damage to your valuable shrubbery.

The foreman on your job will identify potential problem areas before work begins. We thank you for your cooperation.

Joe Doe   *Larry Doe*

Sample operations statement
Figure 3-6

Chapter 4

Getting the Word Out

Finding enough of the right kind of work is the key to success in any painting business. One good way to find the best jobs is to keep your company name in front of the public with effective advertising. Even a limited promotion budget can make the public aware of the superior services you offer.

You may have the most skilled, most conscientious crews in the world and be the most experienced painter in town. But if no one knows, your craftsmen will be the finest, idlest painters anywhere.

I believe that every painting contractor should learn to promote his services. I also believe that most painting contractors are among the world's worst salesmen. So it won't take much to make you the best painting salesman for miles around.

In this chapter we'll be looking at ways to sell your services. From door-to-door flyers to the Yellow Pages, from bulk mailings to the best sales method of all, word of mouth.

Know Your Company

Before you can begin to sell your services, however, you have to know just what those services are. What kind of company do you want to build? What part of the market do you want to work for? How do you want the public to perceive your company? It's time to make some decisions, to focus in on your company's "personality."

The Company Name

If you're just getting started in the painting business, selecting a company name will be the first important decision. Some clients will pick you out of the phone book on the basis of your company name alone. Everyone you deal with has to remember your name. Select the company name carefully to suggest quality and professionalism.

I suggest that you do a little market research before selecting the name. Find out what names appeal to prospective clients in your community. If you plan to work for homeowners, apartment owners, or building contractors, sample those people. This survey can be done in person or over the telephone, but telephoning saves time.

Start by brainstorming as many names as possible. Ask your employees, family and friends for suggestions. Write these names down even if they seem ridiculous. When you have a long list, eliminate all but the five or six you like best. Favor names that suggest quality, your area of specialty or your geographic location. Then check these against names in the Yellow Pages in your area. Eliminate any that are the same or very similar to names already in use. If you will do business as a corporation, be sure the name is not in use by another corporation in your state. Call the office of the Secretary of State to check the availability of the name you have chosen. When you've settled on a name, file a Fictitious Business Name Statement with the County Clerk through your local newspaper.

Now, start the survey. Read the names that made your final list, then ask the question, "If you were hiring a painting contractor and had only the name to go by, which would you select?" Ask this question of at least 100 people. From this you should get at least 50 usable responses. The top selection doesn't have to be your choice, but it merits careful consideration.

Define Your Company
Once you have a company name, the next most basic step is identifying what your company can deliver. This seems simple enough, but it's not. Survey yourself and your employees. Decide exactly what your skills are. Then survey your equipment so you know the kind of jobs you can do best. Put this information down on paper so you can see at a glance the work you want to go after.

Promote only the services you can deliver. As you gain new skills and acquire new equipment, expand the scope of your promotion. Holding yourself out as an expert on high-rise jobs when you've never had one can have disastrous consequences. Bidding work you can't handle can cost you money and destroy your reputation. Learn how to do the work, then promote that work to prospective clients.

Design a Good Logo
A logo is a symbol or design that's associated with your company. It appears on your company letterhead. It's on your business cards, ads, signs, invoices, company documents and promotional literature.

The logo is your company signature. It's the symbol by which you are instantly recognized. Through it your public becomes familiar with you and begins to form an opinion about you.

It should be attractive, creative and original. Later in this chapter you'll find a section on writing advertising copy. All the rules that apply to writing copy apply to your logo, so review that section before you start designing a logo. Figures 4-1 and

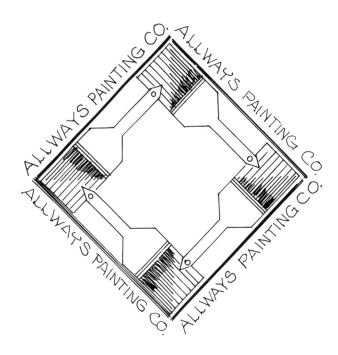

Company logo
Figure 4-1

4-2 illustrate two painting company logos that might lead you to ideas for your own.

Cards and Stationery
You're not a businessman without a business card. It's your calling card. Pass it out to everyone you meet in business. It's the cheapest advertising you can do. Most people in business have a file of calling cards. Your card goes in the file, so it's available when they need a painter.

Company logo used on sign
Figure 4-2

The business card should carry your name, your company name, address, phone number, your logo and your contractor's license number. If you're bonded and insured, mention that fact on the card. But keep the card clean and neat. It has to be easy to read, not cluttered. Your personal name should be featured prominently on the card. This is because most people tend to remember you and your name, not the company name. Figure 4-3 is a sample of a good business card. If you want more help, the printer you choose to print the cards can probably give you some design ideas.

**Business card
Figure 4-3**

The company stationery should have the same information as the business card. The difference is that it doesn't have your personal name. This is so that anyone in the company can use the stationery. Normally, the company name and logo are at the top of the page. All other information is at the bottom of the page for the sake of balance. See Figure 4-4.

Know Your Public

Now that you've established what type of work your company will do, find out what your customers want and need. Again, I recommend a survey. Make up a questionnaire and have about a hundred copies printed. The more people you survey the better, but fifty to one hundred should be enough.

Make Your Survey Effective

The most effective way to do the survey is face-to-face. Of course, that takes a lot of time and may not be practical for you. Second best is a telephone survey. It takes less time but still gives the benefit of personal contact. The last choice is taking the survey by mail. Most people won't take the time to fill it out and return it, even if you provide a stamped return envelope.

Conduct the survey in the area you intend to serve. If you want to paint homes in an upper-income area, that's the place to start. If you want to work for custom home builders, survey them. Each group has distinct needs and requirements.

In addition to finding out exactly what your targeted public needs and wants, your survey should bring out the prejudices and opinions of those surveyed. You'll be dealing with these opinions when you draft promotional materials and letters. Understanding common opinions and prejudices will also help you to understand prospective customers when you meet them for the first time. If you know what someone wants, you're in a better position to provide it. If you know someone's viewpoint, you're more able to appeal to it. Done correctly, your survey becomes a key marketing tool.

Figure 4-5 is a sample survey. It's designed for homeowners, but could be adapted for use with other groups. For example, if you're surveying developers who build large housing tracts, substitute these for the first four questions in our sample:

1) About how many houses do you usually have under construction at the same time?

2) How many housing units do you expect to be painted per day?

3) Do you use union or nonunion painting contractors?

4) Do you want the same crew to do the painting and staining?

Tabulate the Survey Responses

When the survey is complete, tabulate the answers to each question to find what the group wants and needs from a painter. What you're looking for is the majority answer to each question. You'll write promotions to meet the needs and interests of the majority of the people responding.

ALLWAYS PAINTING COMPANY

11458 BURBANK HIGHWAY
PHONE 271-2300
LICENSE #000-321

JUMP OFF, ARIZONA
ZIP 00101
BONDED/INSURED

Company stationery
Figure 4-4

Pacific Painting
Homeowners' Survey

1) How often do you paint your home?

   Exterior:_____

   Interior:_____

2) What time of year do you paint?_____

3) Do you do your own painting? _____

4) What jobs would you hire a painter for? _____

5) What do you expect from a paint job? _____

6) What is your opinion of painters? _____

7) What is your biggest complaint about painters? _____

8) What do you like most about painters? _____

9) Would you sacrifice a little bit of quality for a job done fast?_____

10) What's more important:

    Low cost _____

    Top quality work _____

    Reasonable cost and quality _____

11) What do you most want or need from a painting contractor?_____
    _____

12) How do you select a painting contractor?

    Newspaper ad _____

    Yellow Pages _____

    Friend's referral _____

    Other (specify) _____

Customer survey
Figure 4-5

Here's an example. Suppose the results of the survey indicate that 75% of the people surveyed consider reasonable cost and quality most important, and 80% want the job done fast. Another 60% of the people consider painters to be sloppy and 90% want a neat, good-looking paint job. Tailor your promotion so that it responds to these interests. Your promotion might include: *Professional career painters. Neat and clean! Fast, quality work at a fair and reasonable price.*

Notice that the ad copy reflects the needs, attitudes and views of the group it is designed to reach. This is the key to a successful promotion. Find out what your public wants and then offer what they're looking for. All successful contractors do this. They may not be as well-organized as you are. But they've learned through experience to provide a service that meets customer needs. That's why they became successful. Using a survey in the beginning might have put them on the right track much sooner and with less distress.

Writing Promotional Copy

In advertising, the term *copy* refers to text designed to persuade someone to take a particular action. Your surveys should give you a starting point for writing promotional copy. For example, suppose your survey results show many complaints about painters being late and not completing the job on schedule. Your copy might then read like this: *Joe's Painting — The On-Time Painters. We finish our jobs on schedule!*

After you've decided what basic points your copy will make, it's time to start writing and polishing. Here are some of the basic rules for writing effective advertising copy:

1) Grab Their Attention— You have only a second or two to get someone's attention. That's the time it takes for someone to glance at your promotion. If it doesn't get his attention immediately, most likely it never will.

There are many ways to get someone's attention: it could be a unique and clever logo, a picture, a slogan, or a statement. It could be the use of graphics or color. Whatever it is, if your promotion doesn't have it, effectiveness will be low.

2) Keep it Simple— Promotional copy should be brief and simple. If you're communicating with the general public, the average person must be able to read and understand it quickly.

3) Keep it True— Your copy must be true. False advertising is illegal. Even if it weren't, false advertising is bad for business. Any business built on lies is at a big disadvantage. Lies and dishonesty don't work.

4) Keep it Relevant— Advertising copy must be relevant to the public it's aimed at. If homeowners receive a promotion designed for developers, the results will be predictably dismal. When writing copy, write as though you were talking to the person who will receive the promotion.

5) Make it Attractive— The promotion must be visually attractive. It doesn't have to be a work of art. But the human eye is repelled by images that aren't attractive. Your promotion should be pleasing to look at and to read. The more visually pleasing, the better.

6) Use Your Logo— Your company logo should be prominently displayed on the promotion.

7) Request Immediate Action— The copy should request that the recipient *call now* or mail in the reply card *now*. Always include a request for immediate action. This little extra nudge is all that's required to move many people to action. This seems incredible but it works. Left to their own initiative, many people can put off a paint job indefinitely.

8) Keep it in Good Taste— The copy should not be so harsh or critical that it offends people's sensibilities. Stay within the bounds of good taste and acceptable behavior. Your ad has to reflect your position as a responsible professional in the community.

Figure 4-6 shows a sample flyer that follows these rules of effective copy writing.

Positioning Your Company

Positioning is an advertising term. It means associating your company or service with some other company or product or service. Positive positioning means getting your company compared favorably with another product or service that your public holds in high regard. For example, saying "Joe's Painting is the Rolls Royce of Painters" is positive positioning. Figure 4-7 is a flyer intended to promote positive positioning.

Ad copy
Figure 4-6

But the cost would exceed revenue by a wide margin. There are much better ways to reach the people who need your services. Let's look at the most effective and practical ways.

Personalized Letters

Aside from personal referrals, letters may be the best form of promotion for a painting contractor. Letters are very selective, and they work. They go directly to the people you want to reach, and they will be read.

The letter should be typed and double spaced. The message should be brief and interesting. It should follow all the rules of copy writing. It's O.K. to have the letters printed by a printer. But if possible, sign the letters personally. And be sure the print quality is good. Don't use a copy machine. Some copy machines approach print quality, but a printed piece will nearly always look better.

Positive positioning
Figure 4-7

Negative positioning portrays your company as the opposite of the public's negative image of painters. You would use negative positioning to show you as the good guy. For example, you might say: *Tired of sloppy painters and messy work? Then call Joe's Painting now for professional work and service.*

Promotional Avenues

There are many ways to present your message to the public. Most will cost more than you're willing to spend. And, fortunately, most of the very expensive ways are not appropriate for a painting contractor. Taking a 30-second ad on national TV during the Super Bowl might sell a few paint jobs.

Enclose a business card with each letter. You should also include promotional literature, such as a flyer or brochure. But be sure that these are relevant to the intended audience.

It's strange, but true, that most people have to receive several letters before they will take you seriously. Set up a mailing schedule so that your intended audience gets a letter not more often than once every six weeks. This can vary, and may be influenced by the size of your budget and your mailing list.

A stationery store or print shop will have sheet labels that come with 33 labels to a sheet. Use the grid sheet that comes with the labels, or make your own grid. This is your guide for typing the names and addresses on plain paper. The typed sheet is your master. When you need mailing labels, copy the master list onto the labels. Keep the master in your file for future use.

When your business is small, you'll probably be able to make up your own mailing lists and keep up with the weekly mailings. Eventually, though, you may want to make the leap to bulk mailings, with names from lists you have purchased. Later in the chapter, I've included a section on "Mass Mailings." That's a much more complicated process. But when you've grown enough to need it, the information is there. This type of mailing is aimed at individuals or businesses that you haven't done any business with yet.

Central Files

I keep a list of all the people my company has ever done business with or made an estimate for. I call this our *central file*. The file has a manila folder for each customer. The folders are labeled and filed in alphabetical order by group: homeowner, builder, architect or designer.

Each folder has the name, address and phone number of the client. It also includes the type and date of work done or estimate made, along with any other comments. Here's an example from the John Doe file:

Painted exterior of house on 7/27/84. Price was $1,500. Suggested we repaint interior and gave estimate of $2,200. Back bedroom & laundry room need paint — look bad. Manager for Zumwalt's in men's dept. Wife is Rose. Two kids.

The letter of recommendation from John Doe is also in this file.

We have a separate promotion program for clients in the central file. Never forget the people you've worked for. They're the best single source of future business. Nurture these clients. They should become regulars.

Obviously this list doesn't require frequent mailings. Homeowners don't need your services as often as contractors. Send homeowners a personal letter every year or so. The letter should include a reference to the work you did for them. This makes it more personal, and preserves the favorable memory of a job well done.

Builders, architects and designers need painters on each job, so they should get a letter every three to six months. Every time you send a letter, make a note in the file. If you receive a reply, file that also. Each letter you send should build on the previous letter.

At first, there's no need to go through the files more often than every three to six months. But as your files grow, you may need to go through them once a month, sending a letter to only one section of the files at a time. Mark the folder of the last name in the current mailing with an index card marked "letters stopped here." Attach it to the file with a paper clip. This is the starting point for your next mailing.

It's best not to send an impersonal letter to the people you've worked for. Ideally, you'll send them a personal letter. But as your business and your files grow, this may not be possible. Personal letters take a lot of time. The next best thing is a general letter that has some news about your company, the community, or painting in general. This is like a newsletter. The contents can vary, but be sure to include something that promotes your company. Announce a new service you're offering, or a discount on certain types of painting for a limited time. Figure 4-8 is a typical newsletter sent to business names in the central file.

A newsletter has a life span of one mailing period — you can send it once to each name on the list. If it takes three months to go through your mailing list, then you'll have to put together a new newsletter every three months.

Newsletters can also be used for your mass mailing list. Be sure, though, that it's relevant to those people.

Newspaper Ads

People who look for a painter in the classified section of a local paper are usually more interested in

Paint Contractor's Manual

ALLWAYS PAINTING COMPANY

    Allways Painting is proud to announce that we recently completed painting and staining five custom homes for Smith Contractors, as well as a 20-unit condo project for Palmdale Developers.

    We recently bought a new International Harvester truck and two new spray rigs. Our crews are now able to handle the type of work you need better and faster than ever before.

    Give us a call. We're always glad to do business with you.

Sincerely,

Bill U. Later

11458 BURBANK HIGHWAY
PHONE 271-2300
LICENSE #000-321

JUMP OFF, ARIZONA
ZIP 00101
BONDED/INSURED

Newsletter
Figure 4-8

low price than a good job. But classified advertising is one of the cheapest ways to find prospective customers.

You can place a classified ad over the telephone. The newspaper will help you write the ad, and tell you the best days to run it. To get good ideas for an ad, look through the paper for ads that are repeated regularly. Those are the ones making money. Ads that have appeared continuously over the last three months have been the most successful. Pattern your ad after those.

Give your ad a chance to work. Let it run for six weeks. But it doesn't have to run every day. If it runs only on weekends, then let it run for six weekends.

Classified advertising is relatively cheap. It's probably a good investment until you're established. Even though it doesn't produce high-profit leads, the cost per lead will be low.

Advertising generally takes about six weeks to show good, steady results. This is an average. It can happen faster.

Yellow Pages
Advertising in the Yellow Pages is almost essential for most painting contractors. The only question is how large your ad should be and what the copy should say.

The general public places a certain trust in the Yellow Pages. Anyone not listed there isn't really in business. It's reassuring to your clients to find your name there.

A simple listing of your name and phone number in bold type in the Yellow Pages is very inexpensive. The rates for display ads vary with the size of the ad and the circulation of the book. The bigger your ad and the wider the circulation, the more people are going to see it and the more it will cost. Figure 4-9 shows an effective Yellow Pages ad.

Unless you have money to waste or are in a hurry to become established, start with a small ad. See how it works. An attractive small ad can draw as well as a larger one.

If you live in a large city, you may be tempted to advertise in more than one directory. Don't do it. Place an ad in the book that covers your primary service area. If that works, *then* consider other directories.

For information on Yellow Pages advertising, call your local phone company. They'll help you design and write an ad. The book comes out once a year. Ad space and location are assigned on a first come, first served basis. The deadline for ads is from three to six months before the new directory is issued. If you miss the deadline date, you've missed your chance for an entire year.

You'll be billed monthly for the ad. Failure to pay can result in your phone being disconnected. That's deadly for business.

Yellow Pages ad
Figure 4-9

Trade Magazines
Ads in trade magazines can be effective if you want to reach general contractors in your service area. Many local chapters of the Associated General Contractors and other contractor organizations put out magazines that circulate only to the local chapter. Apartment owners' organizations and building management groups also put out their own magazines.

There are two ways to advertise here: display ads (these run on regular text pages of the magazine) or classified ads. The display ad costs more but is larger and is seen first. Look at Figure 4-10. Advertising managers for the trade magazines will be happy to send a media kit that explains everything you need to know to place an ad.

Trade magazine ad
Figure 4-10

Door-to-Door Leaflets

These are effective when you're working in a neighborhood. The leaflet should explain that you're working in the neighborhood at the Jones house. Invite everyone to come over and take a look at your work. This is very effective in residential neighborhoods.

Painters have been handing out leaflets like this for years. They work. You can stay busy for a month on one block if you pass out enough leaflets. You'll get about a 1% result rate — that's one job for every hundred leaflets. But ten jobs out of a thousand leaflets is effective promotion. If you don't have time to pass leaflets out, hire school kids or a leaflet distribution service.

Figure 4-11 is a sample leaflet.

Job Signs

Everyone is curious about what's going on when they see trucks, workmen and ladders at a house in their neighborhood. Satisfy that curiosity. Put up a sign — at least while you're there. That's another cheap but effective form of advertising. Anyone who's interested can see the quality of your work right on the spot.

The sign should include your company name, logo, license number, telephone number and the words *bonded and insured*. Look back to Figure 4-2 for a good example. Make the sign large enough so that it can be read from the street. A sign measuring 2½ feet high by 5 feet long is a good size. Letters spelling out the company name and the phone number should be at least six inches high. Place the sign a few feet back from the street.

Your Press Book

A *press book* is a package of promotional literature and photos that public relations people compile for publicity purposes. You don't need a public relations agent to deal with the press. But you should collect the same type of information and have it ready to present to prospective clients.

Your press book is a sales tool. Use it to introduce clients to your company. It shows them that you're a competent, established professional in the painting field. Anyone who's read your press

**Sample leaflet
Figure 4-11**

book should be willing to do business with you. In fact, most people will *want* to do business with you if the book is done right.

Keep the press book with you any time you're calling on a new client. The client can look over the book while you're making the estimate.

A large, thin portfolio case available at stationery stores makes a nice press book. It's like a scrapbook but is made for professional use.

Here are some of the things your press book should include:

Biographical data on your company: This section, as well as the rest of the book, should be about 90% photographs. The written text could even be confined to captions below the photos. Include a picture of your first truck, placed next to a shot of your current fleet. Include a picture of your office, even if it's just a spare room in your home. Have pictures of your employees and one of the first jobs your company ever handled. Make this section brief but interesting. It should span the entire history of your company and it should be kept current.

Photographs of completed jobs: This is the largest section of the press book. All photos should be in color. Include a representative picture of every type of work you do. Pictures focusing on a particularly aesthetic room, set of cabinets, or portion of a house are best. Show before and after shots. Show pictures of demanding and intricate detail work. Pictures of jobs in progress and your crew at work should be included as well as unusual jobs you've done.

Letters of recommendation: By all means, include these. They can be placed in a group following the pictures of your work or placed at intervals among the pictures. For example, pictures of a job could be placed next to a letter of recommendation from that job.

Awards and memberships: Certificates, awards and membership insignia go in here, too. Memberships could include the local Chamber of Commerce, Painting and Decorating Contractor's Association, and the Better Business Bureau. Any letters of appreciation for donations to a local charity would also go here.

The following suggestions will help make sure your press book is attractive and effective:

• Keep the contents interesting and neatly arranged.

• Make the text short, interesting and easy to read.

• Make sure pictures are varied, and represent each type of work you do.

• Keep your book current. Replace dated letters of recommendations with newer ones every 12 to 24 months.

• Keep it short and simple. Anyone should be able to read the entire press book in five or ten minutes.

Letters of Recommendation
Solicit a letter of recommendation at the end of each job. It doesn't have to be on letterhead or even typed. Just ask for a little thank-you note indicating that the client is satisfied with the job.

These letters become sales tools for future work. They also have a therapeutic value for the clients who write them. Some clients never seem to realize that a job is finished. In their minds, it's never complete. They will forever be looking at it with a critical eye to find something else that needs to be done.

When someone writes a letter of recommendation, it forces them to evaluate the job as complete. This puts an end to the critical phase and probably cuts off any more calls to you.

Some people are energetic and will type up a very nice letter. Others find it hard to express themselves and will put it off forever. That won't do you any good. Offer to supply the pen and paper if necessary. Then wait patiently while they write the thank-you note. It may seem a bit awkward, but it's better than leaving without a letter. Handwritten notes have a charm that may be lacking in a typewritten letter.

Referrals
For gaining a competitive edge, referrals can't be beat. They give you instant credibility, respect and trust.

People rarely give your name to others until they're asked to. If your customer's friend happens to mention that he's looking for a painter, you'll get the referral. But few clients will look for people to refer to you — at least not on their own. It's up to you to initiate the referral. At the end of each job, after you have the letter of recommendation, ask for three referrals. It's the only way you're going to get them. People aren't going to offer them. Why should they? They're not working for you.

These referrals don't have to be people who need a paint job immediately. But they are worth a call and should go on your mailing list so they get promotions from time to time in the future.

The principle behind this and all other sales methods is simple. You get what you demand. And if you don't demand work, no one is going to shove it under your door. Timid paint contractors are poor salesmen. You have to seek out work aggressively. Be polite, be courteous, but be assertive. People need prodding now and then — especially when it comes to something that can be put off indefinitely.

Mass Mailings

Bulk mailings are the quickest and cheapest way to reach a large but carefully selected audience. If you want to develop a large business volume, you'll probably decide to use mass mailings.

The Mailing List

Suppose you want to mail to all building contractors and interior designers in your service area. Start by finding the right mailing list. The Yellow Pages shows most of the general contractors and interior designers active in your area. But the Yellow Pages doesn't give the name of the principal in each company, or the ZIP code. You could get the ZIP by looking in the National ZIP Code Directory. It's a big two-volume set available at many post offices. But that takes time. And typing the list is also time-consuming.

Another route is to rent the names from a professional mailing list company. The cost will usually be under 10 cents per name for lists of about 5,000 names or more. The compiler can supply the names either on sticky labels or on "Cheshire" label stock, which requires machine application. Be sure to specify "pressure-sensitive" labels if you want to be able to apply them yourself.

Mailing list compilers try to keep their lists current, but they're seldom as accurate as the Yellow Pages. We've compiled our own list from the Yellow Pages for years. But this takes time, so we decided to test a professionally-compiled list for one mailing. We were pleased to discover that the compiled list had four times as many contractors as we could find in the Yellow Pages. We elected not to mail to addresses in lower-income areas where it was hard to get work. The remainder were mailed with "Address Correction Requested". on the envelope, so we could test the accuracy of the list. To our surprise, over 30% were returned as undeliverable! That list was pretty stale. We'll avoid that list compiler in the future.

If you use a professional compiler, ask for a guarantee of deliverability to avoid the mistake we made. Of course, if you're mailing to an "Occupant" list, every letter should be deliverable.

Postage costs for first class mail are high. The cost for bulk rate mail is about half of the first class rate. But bulk mail has special requirements. For example, you'll need a permit from the central post office for your area. There's a one-time application fee and a smaller yearly renewal fee. But this money is quickly earned back in reduced mailing costs.

The advantage of bulk rate mail is the low cost. But there are some disadvantages:

1) All letters must be ZIP coded.

2) Each mailing must have at least a minimum number of letters.

3) The letters must be grouped and bundled by ZIP or SCF (sectional center facility).

4) Bulk mail will take seven days to reach the addressee.

5) If an address is incorrect, your letter won't be returned or forwarded unless you request an address correction. This will cost you 25 cents per correction.

Bulk mailing requires more time and effort. But if the volume is high, the savings are worth the trouble. If you decide to try bulk mail, the first step is to get in touch with your local post office. Make sure you understand *all* of the regulations before you start, or you may find yourself at the post office with a thousand unmailable letters.

Organize Your Lists

Keep your mailing lists in a file drawer in manila folders identified by group. For example, building contractors might be in the first folder, architects in the next, and interior designers in the last.

Your mailings would start from the front of the first folder and proceed toward the back. Use a marker to indicate where the last mailing ended. This could be a slip of card stock with the words "mailing stopped here" printed on it. Attached to a label sheet with a paper clip, this indicates where to start the next mailing.

The volume of mailing depends on your budget and your marketing goals. But don't exceed one mailing to a prospect in any six-week period. If they hear from you more frequently than that, prospects begin to think of you as a fanatic.

More important than volume is the consistency of mailing. Keep up a steady flow of promotion. Set a goal of so many pieces a week. Stick to that standard. Don't skip mailing for three weeks and then catch up in the following week. You want a steady flow of jobs, estimates and inquiries. Of course, the more letters you send out, the more responses you'll see.

The Time Lag

Expect a six-week delay between the time a batch of letters goes out and the time inquiries start coming in volume. Figure 4-12 is a chart showing the correlation between the number of letters mailed and the number of estimates requested during a six-week period. You'll start getting phone calls a week or two after you've mailed a batch, but a strong, steady flow will take about six weeks. This is why it's so important to mail consistently, on a regular schedule. Don't wait until you're desperate for business. Results will never be immediate.

This is a fundamental principle of promotion. All forms of promotion have a time lag between the date they start and the date they bring in strong results. The lag will usually be at least six weeks, no matter what form the promotion takes. There are, of course, exceptions to this rule. But you won't succeed in business if you base it on hopes or exceptions.

The Promotional Material

Except for the newsletter, most types of promotional material have a fairly long life span. In practice, you can use the same promotional piece as

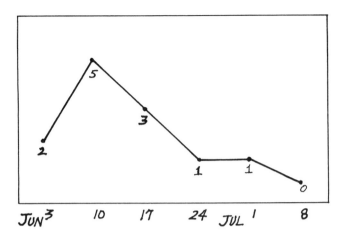

A) Estimates made before promotion started

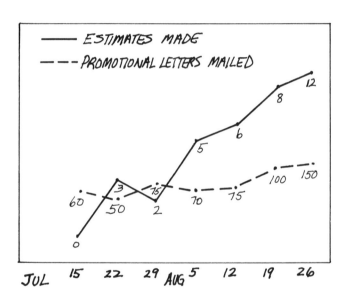

B) Estimates made during promotion

**Promotion results
Figure 4-12**

long as it brings in satisfactory results. If you have a flyer that brings in a flood of phone calls every time you mail it out, don't change it until the action drops off significantly.

In any promotion, keep doing what works. Once you find a successful vehicle, stick with it. If it's unsuccessful, drop it. But dropping something that works is like cutting your own throat.

Quality vs. Quantity

Every promotion must strike a balance between quality and quantity. Don't fuss over a letter or flyer for weeks trying to make it perfect. Your business could go down the drain in the meantime. And don't create a mailing piece so expensive that you can't afford to send it to more than a handful of people.

My advice is to keep the cost low. Find out what type of flyers and offers other painters are using. Something that's used over and over again is probably working. Adapt it to your own use. Keep mailing that piece as long as it pays off. Then make small changes. If that improves results, keep mailing. If it doesn't, try something else.

Promotion and Gross Income

For most paint contractors, the money you spend on promotion is proportionate to business volume. The more money you spend on advertising, the more work you'll have. If you're not spending at least 2% of every dollar of gross receipts on promotion, you're under-promoting. Increase your promotion budget and watch sales climb.

To prove to yourself that promotion brings in sales, keep a chart of the number of letters mailed out each week. Also chart your gross income weekly. You'll find that sales follow promotion. As the number of letters increases, gross income increases. Six weeks after you start mailing, the gross income chart should be climbing. See Figure 4-13.

But don't think that merely mailing letters will make you rich. Everything else has to be working too. But if your promotions work, putting out a steady stream week after week will keep a steady stream of work coming in. That can result in a very comfortable income.

Be Persistent — It Pays

Here are some final notes on promotion, based on my experience over the years:

Don't give up too soon— Never stop sending promotional material to a past client just because he doesn't respond. He may be doing business with another painting contractor. But that relationship could end at any time. When it does, your promotion should be there. Don't give up too soon on any promotion. People file away an ad, then dig it out again when they need a painter. I once had a client call me and quote a special offer from a letter

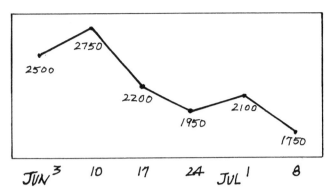

A) Gross income before promotion started

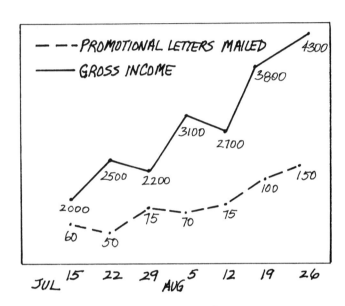

B) Gross income during promotion

**Promotion and gross income
Figure 4-13**

I had sent out four years earlier!

Here's a curious fact about promotion. You can send a promotional piece to one group of clients and get no response. But suddenly you'll start getting jobs from a different group of clients. This may seem odd, but I've seen it happen many times. So don't let one failure discourage you. Just keep promoting. Send out enough promotional pieces and you'll get responses . . . sometimes from the most unexpected sources.

Keep promoting even when you're busy— Promotion should be a steady, continuous process. When you're booked solid and it seems that you'll never run out of work, keep promoting. Abundance lets you be selective about the work you take on. Raise your prices a little to control the work flow. Abundance won't last forever. And it didn't just happen. It was created by your own hard work. To keep that abundance, keep promoting.

Once you've let your promotion drop off for several weeks or months, it takes six weeks to get started again. So never stop promoting. Neglecting sales is leaving the future of your business to chance. And that's inviting disaster.

A final word. When you're first starting out, and even later, when you're prospering and have some spare time, go out and knock on some doors. Meet prospective clients in person. Take your press book and samples (such as stainwork, antiquing, etc.) along. Let these people know that you *want* their business. Be enthusiastic.

Call ahead and set up an appointment if possible. If not, just go out there and have at it.

Nothing, absolutely nothing, is better than a personal one-to-one meeting with your prospective clients.

Chapter 5

Introduction to Estimating

Every paint contractor has to be a paint estimator. Successful paint contractors know how to develop cost estimates that are both competitive and profitable. This is vital to success. Even if you do everything else right, if you make bad estimates, you'll fall flat on your face.

Bad estimates come in two varieties: too high and too low. A bid too high loses the job. A bid too low loses your shirt. You can survive high bids. You'll just have a lot of spare time. If your prices are too low, you'll get plenty of work. But pretty soon there won't be enough money to make payroll or buy materials.

In this chapter I'll describe my method for making accurate and profitable estimates. First, I'll make some general comments that apply to nearly every estimate you make. Then I'll show you how to estimate repainting existing buildings and how to estimate the painting of new construction. Finally, I've included a sample plan and a sample estimate based on that plan.

Learning the Hard Way

Like most estimators, I've learned what I know through trial and error. I jumped in head first with no prior experience. Every time I made a bad estimate, I learned something. I learned to avoid a mistake like that in the future! Once you've made every mistake possible, you're an experienced estimator. But that's an expensive and slow way to learn estimating. I wouldn't recommend it to anyone.

Let me save you the money and time I wasted. This chapter has the information I needed when *I* started estimating. Use the procedures and suggestions here to avoid the headaches, losses and frustrations inexperienced estimators usually have to endure.

The Basics of Estimating

After ten years of estimating all types of work, I can walk on a job and have a good idea of the bid price before doing any calculations. But this "eyeball estimate" is seldom my final bid. On anything more than the most routine and simple job, I make a complete labor and material cost estimate. If my first guess was within 10% of the cost estimate, I feel confident that my estimate is correct.

You'll probably develop the same ability to judge the cost of many jobs on sight. But I hope you'll have the wisdom to realize that there's no magic way to estimate painting costs. You have to go through the steps I'm going to describe in this chapter. Anything less is going to produce haphazard results at best.

The next thing to understand about estimating is that there is no single "right" price. I've seen jobs where the estimates ranged from $20,000 to $60,000. Don't be too sure that one or both of these bids were wrong. It's possible that both the high and low bid were correct for the contractors that submitted them. Every bid has to be custom-made for the crew and equipment that will do the work and for the contractor that submits the bid. Don't think of estimating as a treasure hunt — everyone trying to find the right price. There isn't any single right price. There are just good estimates and bad estimates — *for the contractors who submit them.*

Contractors tend to estimate according to need. That is, if they have plenty of work lined up, they raise prices to control volume. If they're short of work, they shave prices a little. I'm not going to condemn that. It's human nature and it's probably good business, too.

But there are some guys who will work for practically nothing. You'll never underbid these characters. I *will* condemn these cutthroats — not for their low prices, but for the slipshod job they usually do and the bad name their careless work gives to the painting profession.

Don't try to underbid lowball painters. You do a different class of work. Let the cutthroats go their own way. They won't last long. When they're gone, you'll still be in business doing professional quality work at fair prices.

Here's the point you should have picked up so far: You can't base your prices on what others charge. There is no reference book, no competitor, no other estimator, no sage of the painting industry, that can determine what your prices should be. That's for you alone to decide. And you determine that by building estimates step by step, as I'm going to explain in this chapter.

What are the steps in estimating? Well, I'll tell you what an old paint estimator once told me. "It's easy," he said. "Just figure out what the materials will cost, how long it will take to do the job, how much you want to charge for your labor. Then add up the total. Aside from that, there's nothing to it."

Put very simply, that *is* all there is to it. But of course there are a few details to consider. It's the details that fill the remainder of this chapter.

As you read this chapter, keep in mind one important concept: *The estimated cost of something you forgot is always zero.* Forgetting something is the cardinal sin in estimating. You can underestimate labor by 10 or 15%. You can allow for too much paint. Mistakes like that won't hurt too much. At worst, the bid might be 5 or 10% too high or too low. If you're lucky, the mistakes will cancel out so the actual cost is very close to the estimated cost.

But forget one important item and your profit is gone. Forget to include the prime coat, omit a bedroom, leave out the cost of scaffolding and you've got a sure loser. That's why I recommend careful, step-by-step estimates. Follow the actual painting sequence as you prepare the estimate. Visualize each part of the job. Figure the labor and material for that part. Then move on to the next portion. Don't take any shortcuts. Put every step down on paper. Then be sure there's a labor and material cost listed for each step.

If there is a trick to estimating, it's this. *Identify every cost item and estimate the cost of every item you identify.* The most successful estimators are always the most thorough. They find every cost item and put a number down beside it. That figure may be too high or too low. But at least it's an estimated cost. It's never a 100% underestimate.

Once all your figures are listed, it's a simple matter to add up all the numbers.

Experience is the Best Teacher
Paint contracting is a highly competitive business. You get most jobs by being the lowest responsible bidder. Every bid is a bet, a wager that you can do the work profitably at a set price. It takes experience to win in a game like that.

On routine work, most estimators are on pretty safe ground. But on more complex jobs, opinions (and estimates) can vary widely. If you're estimating a job you don't understand, either get help or pass up the work entirely. You can usually find an old contractor who's handled the type of work in question. Only someone who's done the work can know what it takes to get the job done, what type of special equipment you'll need, what problems you can expect and how to handle them. If you find someone who has the experience and knowledge to estimate a complex job, offer to pay an estimating fee of 3 to 5% if you get the job and make money on it.

I follow two rules when estimating an unusual job. First, even for the best estimators, 10% of all jobs are losers. Unless you're a magician, it's better to pass up some work. Pick and choose what you handle. Specialize in profitable work, not the 10% that no one could turn into a winner. Second, when a job you want involves unknowns, allow enough margin in the estimate to cover several mistakes. Take a flyer occasionally. But be sure there's enough potential reward to make the risk worthwhile.

Learning from Mistakes
Novice estimators are seldom good estimators. It takes practice to be good. When you make a mistake, make the most of it. Keep good records on every job you handle. When a job is complete, note the actual cost beside the estimated cost for each major item. Compare the estimated and actual costs. If there's a big difference, find out what went wrong.

Write a little summary of what you learned on the job and file the estimate and summary for future reference. Keep these estimates in your filing cabinet. When estimating a new job, review the record of a similar completed job. Get the full benefit from what you learned. Identify mistakes to avoid repeating them.

Even estimates several years old can be valuable. The hourly labor cost and the cost per gallon of paint have probably changed. But the manhours per 100 square feet for preparation and painting and the coverage of materials will still be about the same. *Cost Records for Construction Estimating,* a book published by Craftsman Book Company, 6058 Corte del Cedro, Carlsbad, CA 92008, provides full instructions and forms for maintaining and using records from past jobs.

Custom Jobs
I use the term *custom job* to identify a better class of work, one that requires detailed painting and high craftsmanship. Usually these are residential interiors. But some offices and store interiors are custom jobs.

Unless you're desperate for work, charge top dollar for custom work. The client expects quality and is willing to pay for your best efforts. Be prepared to spend more time on this type of job and bid it accordingly. No one's going to pay more than the bid price just because you underbid. And no one is going to lower his expectations simply because you're losing money. It's better to charge top dollar and lose the job than to lose your shirt. Money lost is bad enough. But it's also demoralizing to work for less than you're worth.

So, if they don't go for your price, let the job go. You may be without work for a couple of days, but that's better than being saddled with a losing proposition that lasts for weeks.

On a custom job, add at least 10% to the total job cost before submitting the bid. That's the minimum needed to cover the extra time spent meeting client demands.

Estimates to Waste
Chapter 4 explained how to attract requests for estimates. It should be obvious why that's important. No painting contractor gets every job he bids. But generally, the more jobs you bid, the more work you'll have. As I explained back in Chapter 1, increased bidding volume is usually followed by increased workload.

There's no set percentage of bids that you should get. But if you're getting more than one job in three, your prices are probably too low. If you're getting less than one job for each six or eight estimates, there may be too much competition for anyone to operate profitably. And it's certain that you're wasting a lot of time making estimates.

The importance of making a large number of estimates is that it gives you estimates to waste — jobs you would take at your price but don't *have* to get. Making lots of estimates is a security blanket. It avoids the anxiety of having no work.

But a high demand for estimates is not, by itself, a reason to increase your prices. Increase prices when you handle better-quality work or when you're so booked up that you have to limit new work in fairness to existing customers.

When no one's requesting estimates, there's a lot of pressure to lower prices. In the business this is called "bidding it to get it." The rationale here is that any work is better than no work. That may or may not be true. But everyone can agree that it's a bad position to be in. To avoid it, keep promoting.

Doing lots of estimates puts pressure on you as an estimator. But heavy bidding usually means prices you can live with. And occasionally you can even waste one — submitting a *courtesy bid* that's at least 25% too high. That's better than not having any work or taking a job at fire sale prices.

The Customers You Bid For
Everyone has heard stories about contractors who cheat their clients. It's happened. I won't deny it.

But some clients are pretty good at cheating contractors. And contractors sometimes cheat subcontractors. You won't see stories like that in the newspaper. Apparently there's little sympathy for contractors and subcontractors that get cheated. I guess contractors and subs are supposed to be able to take care of themselves.

When you spot a cheat, there's only one way to handle it. Complete whatever work you have to do and then keep your distance. No team of lawyers, no arbitrator, not even the police or sheriff can salvage the deal if your client is determined to cheat you. And don't expect that you can outfox an experienced cheat. There's no way!

I've been cheated a few times. The job I described in Chapter 1 is an example. As I look back, I can remember having doubts when I first met those people. My initial reaction was, "There's something not quite right about this guy." In each case I had an intuitive feeling that everything wasn't on the up-and-up. A couple of guys simply would not pay their bill. And some paid but took so long and caused so much aggravation that I lost on the deal.

So I now have an ironclad policy: I don't do business with anyone who gives me reason to distrust him. I may pass up a few good jobs. But I've passed up more than a few headaches. Making lots of estimates lets me be selective, not only about the work I take, but also about the clients I work for. When I'm making an estimate, I'm also evaluating the client who's going to pay the bill. If either the client or the job doesn't meet my standards, there's other work to do.

There are two reasons for someone to hold out on you. Either they intended to cheat you right from the beginning or they became upset with you for some reason. A little touch-up or repair work will usually win over an upset client. A deliberate cheat intends to hold out as long as possible.

Time and Material Estimates
Most contractors see selling jobs on a "time and materials" basis as good work. True, it simplifies estimating. There is none! It also takes all the risk out of the job. In theory, there's no way to lose money. Your client pays for all materials and agrees to pay a set cost per manhour for all labor. If you have a small markup on materials and price the labor correctly, it seems that the job should be a sure money-maker. And there won't be many complaints or requests for touch-up when the client's paying for the extra time!

Unfortunately, time and material pricing doesn't always work out the way it should. First, owners feel like they're writing the contractor a blank check for any materials he wants and for any number of manhours he can waste. Unless the client has absolute faith that the contractor is totally honest and trustworthy, the client probably feels safer with a set price. And every paint contractor who's used time and material pricing on more than a few jobs has probably had a dispute or two over collection. Few homeowners understand the complexity of painting. Present the time and material bill for collection and your client may launch into an explanation of how long the job should *really* have taken. The client will argue that he shouldn't have to pay for mistakes and do-overs, even if he caused the problems.

Occasionally I do work on a time and material basis. Usually these are either very small jobs or involve an unusual request by the owner that makes it impossible to estimate costs before work begins. If the owner insists on supervising the job or demands the right to approve surface preparation before work begins, I suggest that we work on a time and material basis. Usually this provides enough motivation for the owner to approve my original fixed price bid.

Estimating Steps
Before we actually begin compiling labor and material costs, let's review the estimating process. Here are the steps you'll follow in making an estimate:

1) Contact the customer and set up a time for making the estimate.

2) Arrive on time.

3) Go over the job with the customer.

4) If necessary, walk through the job a second time by yourself. Take all the time you need.

5) Write down your notes and figures on the estimating forms as you go along. Refer to your estimating checklist and manhour requirements table whenever necessary.

6) If possible, total the estimate and give the bid to the client on the spot. If more time is needed, tell the client when to expect the bid.

7) Get the estimate to the client on time. If he's in a hurry, offer to call in the estimate total and then mail the written estimate to him. If a delay is unavoidable, notify the client before the date the estimate was due.

8) When the client agrees to your bid, either get a signature on your bid or write up a contract and get it signed as soon as possible. This commits the customer to having you do the work. It also prevents another contractor from taking the job away from you.

Delivering the Estimate
It's just good salesmanship to get the completed estimate into the hands of your client as soon as possible. On a small job, I can write up the estimate right on the site. Carry your company stationery and some carbon paper in your truck. Compile the estimate on a contract form and offer it to your prospective client for an immediate decision. If your client decides while you're standing there, you'll get the job nine times out of ten. On a more detailed or bigger job, phone the cost in the next day and follow that up by mailing in the written estimate.

Getting the estimate to the prospective client as soon as possible shows that you're professional and capable. That's an important step in building respect.

Here's the important point: Requests for estimates are like fresh fish — after two or three days, they're better disposed of than saved.

Starting Out
For your first estimates, start with smaller jobs and work you know. Until you have more experience, limit the risk as much as possible. In a one- or two-man company, a couple of bad estimates can be fatal. Don't take on unfamiliar work if the margin for error is too slim.

If you have no experience in pricing, find an experienced paint contractor who's willing to be candid on the subject. He'll be able to suggest a range of accepted prices. Ask other paint contractors and general contractors you know. Some will talk to you. Others will feel that giving out their usual prices is like giving out the combination to their safe.

After completing a few estimates, it's easier to develop a "feel" for prices in your area. Any time you submit an estimate but don't get the job, contact the general contractor or homeowner and ask him what the job went for. If he's reluctant to tell you, remind him that you made the estimate at no charge. He owes you the courtesy of revealing the winning bid. You're not asking for the name of the contractor who won the bid, just the amount. Use this information to adjust future estimates.

It's common for a business just starting out to undercut prices quoted by the competition. That tends to attract customers and develop a broader clientele more quickly. I suggest you do the same, within reasonable limits. But there are two things to remember if you follow my advice.

First, set your prices low enough so that you do attract business. About 10% below the going rate will usually be enough discount to attract price-conscious clients. But don't set prices so low that you can't survive. There's no advantage to working for nothing.

Second, don't put any faith in a customer's promise to refer lots of work from friends and relations. If you quote an attractive price and do good work, that's incentive enough for others to call. Promises of referrals are worth little or nothing. They're offered by chiselers trying to beat down your price.

As you become better established, gradually raise your prices to where they should be.

Raising Prices
There are two situations that call for higher prices. One is when you go from doing production-type work to custom work. The quality required in custom work takes longer. Your prices have to reflect that change.

The second time to raise prices is when demand exceeds capacity. You could either increase capacity by hiring extra painters or control volume with higher prices. Slightly higher prices will still land enough work to keep you busy. But raise them too high and you'll run low on work fairly soon.

If you decide to raise prices, watch how bidding volume and contract volume change. If you usually get half of all the work you bid, a 10% price increase should still leave you with about 40% of the work bid. Watch this percentage carefully. If you get less than 25% of the work bid, ease back on those price increases. If volume of work doesn't drop at all, another 10% increase might be in order. Then stay at this price level until you get another large increase in demand.

This strategy for price increases takes into consideration both the demand for your work and the value the public places on your service. But notice that these increases are not the same as price increases based on changes in the cost of material and labor. Changes in labor and material costs affect all painters about the same and can usually be passed on to your customers immediately.

Consider price increases based on higher volume a reward for your hard work and quality service.

Bidding for Interior Designers
Bids prepared for interior designers should be 10% higher than similar bids for homeowners or general contractors. That 10% covers the hours you'll spend answering questions, resolving complaints, mixing colors and painting samples on the wall.

Here's something to remember when working with interior designers. Many designers want cut prices from the painters they deal with. A discount from the paint contractor leaves a better margin for the designer. But unless you're running heavy volume with a particular designer, I don't see any reason to give a designer lower prices than anyone else. In fact, I've found that designers are often difficult clients and should pay more.

Remember also that a designer is working for his client. In any dispute over your work, the designer will always side with the owner. That leaves you alone in defending the work you've done and the legitimacy of your bill. If you have to sue to collect, the designer will try to slip out of the middle, forcing you to sue an owner you've never dealt with directly. And the owner may claim that his only responsibility is to the designer — and the designer never did what was promised!

That's why my prices for designers are 10% higher.

List Prices
As I said earlier, there's no such thing as uniform prices for painting. But there are generally-accepted price standards for different types of work. This is more true in new construction, where high production is the goal, than in custom painting, where quality and personal satisfaction are of primary importance.

Most of the paint contractors I know estimate the basic cost of interior painting in new residential construction by the square foot of floor area. This generally works because the interior of most new residential construction is remarkably similar from a painter's standpoint.

If you establish a price list to make estimating easier, it might look something like this:

Condominiums, Apartments, Tract Houses
— One coat: $.75 per square foot
— Two coats: $.90 per square foot
— Three coats: $1.05 per square foot
— Add 20% if the ceilings are painted
— Deduct 20% if the cabinets are prefinished

Spec Houses and Single Homes
— One coat: $1.00 per square foot
— Two coats: $1.30 per square foot
— Three coats: $1.60 per square foot
— Add 20% if the ceilings are painted
— Deduct 20% if the cabinets are prefinished
— Add 25% to 50% for custom-quality work

These are base prices. Anything out of the ordinary will increase the cost. When you run into a job that's out of the ordinary, adjust the price accordingly. For example, if the job has a lot of decorative wood trim on the interior, add the cost of painting trim after you've computed the usual square foot price. If the job requires more work overall, estimate the additional work needed. If it's 10% more, increase your square foot price by 10%.

The prices listed above are only examples. They aren't intended to be accurate. But the list you develop may be similar except for the dollar amounts. The dollar costs have to be based on your experience and the competitive conditions in your business area.

If you have a new residential job that's larger, more complex, or is a little unusual, there's no way to estimate it other than to break the work down into its components: each room, each coat, and all the details that have to be handled separately. Then add up the total and compare that with the price you got by using the standard price list.

On a more difficult condominium or apartment building, figure the cost of each unit — the number of rooms, the coats in each room, the linear feet or square feet of detail work. Then multiply by the number of units and add the cost of exterior work — rain gutter, decorative trim and storage cabinets in the garage. Compare this step-by-step estimate with your square foot price for the same building. Then decide which estimate is best for the job.

Painting the interior of most apartments is high production work that makes standard pricing more appropriate. Prices are usually based on the number of bedrooms. To find the going rates in your area, look in the classified section of your Sunday newspaper or a local magazine for apartment owners. You'll find these magazines in the reception area of apartment management company offices. You'll probably find a list of current painting prices there also.

Bidding residential, commercial and apartment *exteriors* is more difficult. Normally, you'll want to break the job down into parts and then find the labor and material cost for each part.

Estimating Tips
Here are some points to observe when estimating any job:

1) Never rush an estimate. Spend as much time as necessary to get an accurate figure. Remember that the estimated cost on everything you forget is always zero.

Don't let anyone else rush the estimate, either. Occasionally, you'll have an owner that wants a snap estimate or "ballpark" figure in the first five minutes you're on the site. They want you to quote a low figure off the top of your head. They plan to hold you to that figure later if the complete estimate comes in at a higher cost. Don't do it. Quote them a figure that you know is high. Then, when you give them your normal price later on, it will seem like a bargain.

2) If you have doubts about your estimate, don't make a decision immediately. Take the estimate back to your shop or home. Talk it over with your partner or another contractor. Sleep on it. Make your decision the next day. You'll usually view the job in a different light the following day. Your judgement has matured in the meantime.

3) *Inspect* all surfaces to be painted. Don't just stand in the middle of the room and look. Run your hand over the walls. Look for hairline cracks.

They have a tendency to appear everywhere when you start painting. Examine *all* the windows and doors in detail. If the room is dark, turn on the lights and open the curtains.

The cost of surface preparation is the biggest variable in painting. Do yourself a big favor. Inspect all surfaces like your profit depended on it — because it does. Describe in the estimate exactly what surface preparation is needed and how you plan to do that work. Then make those points clear in your proposal and contract. You may want to set an hourly rate for any additional preparation that the owner may request. That's fair to both you and your client.

I'll explain how to estimate surface preparation in detail in the next section. For now, just remember how important surface prep is in your estimate.

4) When estimating residential work, never ask an owner what kind of job he wants. That's unprofessional. The owner will always say he wants top quality. But if you bid a truly top quality job, the owner won't be willing to pay. Few residential jobs require the type of quality you would expect in an opera house or a grand ballroom. *That's* top quality.

A professional always does a quality of work that's appropriate for the building and the owner. Discuss the quality of the existing paint job. If the owner has no complaint about the prior paint job, you can consider that the existing standard of quality is acceptable. If the owner complains that the existing enamel has streaks, make note of this. Include in your estimate a charge for sanding the existing enamel so you can lay off a smooth coat over it.

5) Even if houses tend to be similar from a painter's standpoint, all clients are different. Consider the client and the client's expectations when making the estimate. Get the owner's agreement on who will protect trees and shrubs, who's going to move furnishings, and who's going to protect adjacent surfaces. If the house is filled to overflowing with books, furniture and toys and if the owner expects you to move them, charge accordingly. But don't offer to move delicate items of sentimental value or expensive antiques. That's a no-win situation.

6) Consider safety when making the estimate. Time spent making the job safe is time well spent. But it's also time you have to charge for. If small children will be present while the crew is working, tools and materials can't be left unattended. You'll have to keep paint and thinner in sealed containers and stored safely out of reach. Plan to spend some time on continuous cleanup during the course of the job.

7) Sometimes a homeowner will want you to follow an unusual order when painting his house. For example, I once had a request to paint the kitchen first, then the upstairs bedroom, then the living room downstairs, then the exterior and then the rest of the interior. A schedule like that requires more labor. It's fair to charge for it.

8) If you're going to use more than two or three colors on the same job, charge extra for the time you'll spend cleaning brushes and rollers. Switching colors interrupts the momentum of work and slows production. Take this into account and charge accordingly.

Estimating Preparation Work
The hardest part of most estimates will be figuring the cost of surface preparation. First you have to determine what's needed. Then you have to figure out how long that will take. Making these decisions requires knowledge, experience, and careful observation.

No two surfaces will be exactly alike in an older home. Every door, window and wall to be painted is a special case. One door may need fifteen minutes of scraping, patching, and sanding before it's ready for paint. Another door in the same room may need only a light sanding.

When you estimate prep work, examine every surface. Move curtains back away from windows. See how difficult it will be to remove dirt and grease that's accumulated on the walls. Open doors and windows and examine all edges. Some windows may be painted shut or inoperable. Freeing a stuck window takes time and may result in damage that has to be repaired at your expense.

Look over every wall, the entire ceiling and all doors and windows. Pay special attention to areas that show water damage or deterioration. Exclude from your estimate the cost of repairs such as stripping off wallboard and replacing studs. Use a nail or screwdriver to probe into deteriorated wood, plaster or drywall. If the deterioration extends into the surface more than a few hundredths of an inch, extensive repairs may be needed. Point this out and exclude the work from your bid. If the owner insists on painting over deteriorated surfaces without adequate preparation, make it clear that you take no responsibility for paint durability in those areas.

Always look for chalky paint when you're estimating surface prep. Chalky paint is paint that has decomposed from exposure to the weather. If the paint is severely chalky, it may require extra preparation work and an additional coat of paint. To test for chalkiness, rub your hands over the surface. Chalky paint will leave a white residue on your skin. If the paint is severely decomposed, your hand will look like it's been erasing a chalkboard. If the paint is only moderately chalky, adding a bonding agent to the paint may be enough. But all types of paint, on any surface, are subject to chalkiness. Paint applied to a chalky surface will not bond properly. Don't make this mistake.

Using the Checklists
Most experienced estimators use checklists when estimating. A good checklist serves as a reminder system — suggesting things you may have overlooked or omitted. If you don't have an estimating checklist, use mine. My interior and exterior checklists are at the end of this chapter. See pages 123 to 134. If you already have a checklist, borrow any parts of mine that you like better than the checklist you're now using.

As you walk through the job to be estimated, check off on your checklists the work to be done. Note the dimensions on the form. Fill in as much of the information as you can. Consider carefully the entire job from beginning to end. For each item on the checklist, picture in your mind exactly what has to be done, and in what order. When you've filled out every applicable portion of the checklists, you've captured the important facts required to make the estimate. The next step will be to transfer this information to your estimating sheet and begin compiling prices.

Pricing Materials
This is the easiest part of estimating. You know how much surface area a can of paint will cover. Figuring the number of cans needed is simple. Let's look at an example.

You're repainting a wall. The label on the paint can says one gallon will cover 400 square feet of wall. If you have 1,900 square feet of wall, divide 1,900 by 400 to get 4.75 gallons of paint. Round that up to 5 gallons. If the paint costs $15 a gallon, the material cost will be 5 times $15, or $75.

Of course, that paint costs more than $75 when the time spent ordering it and picking it up are included. Even if the paint store delivers to the job at no cost, someone still had to phone in the order. And someone will have to pay the store's bill at month end. You can see that there's a labor cost in every gallon of paint — even before the can is opened. That's why most paint estimators mark up their material cost by at least 10%. Your cost of getting the paint to the job may be more. But don't mark up material by less than 10%.

Some paints don't list coverage figures. And on some surfaces the coverage will vary. If you're working with a new product or an unusual surface, get the manufacturer or dealer to suggest how much paint will be required.

Estimating Labor
Estimating labor is always a guess. How long will it take to get the job done? No two jobs are exactly alike. No two crews are exactly alike. Even the same crew can have good days and bad days. Weather, size of the job, suitability of tools and materials, access problems and many other factors can change productivity. Estimating the labor required always involves judgement. That's why the best estimators are always experienced estimators.

Having explained that there are no manhour figures that fit all jobs, I'll offer the manhour estimates I use. See Figure 5-35. If you have no other figures, feel free to use mine.

The estimates in Figure 5-35 cover only productive labor. No supervision is included. These estimates assume that the painters are experienced craftsmen. But all figures exclude time for setup, preparation and cleanup, except as noted. Larger jobs might go faster and smaller jobs may take longer than the estimates indicate. Specialized crews working under the direct supervision of the contractor or a highly-motivated supervisor will produce more work per hour than these standards. Even a skilled craftsman will take longer on top quality work.

If the materials needed are not readily available, if temperature and working conditions hinder production, if the tools and equipment are not in good condition and appropriate for the job, expect productivity to be below the standard in Figure 5-35.

But use Figure 5-35 only as a last resort. The best guide to manhours on your next job will always be your record of manhours on your last job. How long did it take to paint a door, a sash window, a 16' high ceiling, an 8' by 10' wall? If the next job is like your last job, the times should be similar.

The figures you compile yourself will always be better than any you find in a book. Keep track of how long your crews spend painting each type of surface, with each type of coating. Good manhour records may be your most valuable intangible asset.

Even if you have a good guess on the manhours, something unexpected could throw the whole schedule off. For example, on a residential interior you may have to move furniture, protect carpeting and work around a house full of kids. On an exterior job you may have to protect shrubbery that takes on extraordinary value the minute the owner sees a ladder approaching. You might discover that the roof is much steeper than you thought. In both new construction and remodeling, some work has to be done in stages because of delays or interference caused by other trades.

There's no way to take every possibility into account when estimating costs. But you can anticipate most of the more common problems. That's why estimating is an art. If it were a science, computers would be doing all the estimating and all estimates would produce the same bid price. We'll never see that day.

Be sure your manhour estimates include the time needed to get set up, the time spent switching from one phase of the job to another, and time spent cleaning up your equipment, shop and work area. At least 10% of crew time will be spent on setup, moving from one part of the project to another, and cleanup.

Manhours for Residential Work
Here are some time estimates you'll use on nearly every residential job. A journeyman shouldn't take longer than the times listed here unless you're doing high-quality custom work. On custom jobs allow 25 to 50% more time than listed because more time is spent on fine details.

Times are for one coat of paint only and do not include any preparation, setup or cleanup. If a second coat of paint is required, that coat will usually take 25 to 30% less time than the first coat. This is because the first coat seals the surface, making it easier to apply the second coat.

These figures assume that work is done with painters standing on the floor. If ladders or scaffolding are needed, allow more time.

Sash windows— The most common sash window has 12 lites. See Figure 5-1. Painting time for this window, frame and sill should not exceed 30 minutes. If the window has more than 12 lites, add two minutes painting time for each additional lite. Subtract two minutes for each lite less than 12.

If the window has only a sash running across the middle of the opening, as in Figure 5-2, estimate the painting time at 20 minutes. That's your minimum time for any sash window. Don't estimate less than 20 minutes for any sash window.

Common sash window with 12 lites
Figure 5-1

Common sash window with center sash
Figure 5-2

Doors— Doors with inset panels will usually have four, six, or eight panels. See A, B and C in Figure 5-3. Four panels per door is most common. Painting time for one side of the door and one-half of the door jamb should not exceed 30 minutes.

Flush doors (Figure 5-3 D) have a flat surface. Painting time should not exceed 20 minutes per side, including one-half the door jamb.

French doors (Figure 5-3 E) usually have 15 lites. A journeyman should paint one side of a French door (and one-half of the jamb) in 40 minutes.

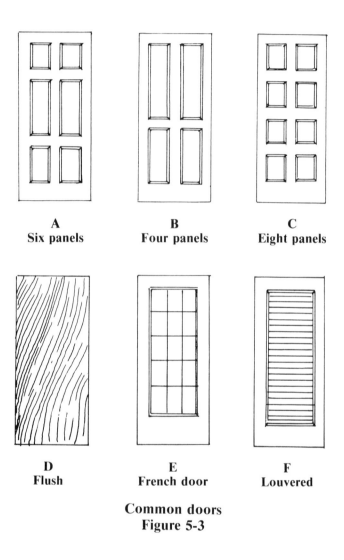

A Six panels
B Four panels
C Eight panels
D Flush
E French door
F Louvered

Common doors
Figure 5-3

Louvered doors (Figure 5-3 F) of a standard size should take 45 minutes to paint, including half the jamb.

Molding— If base molding, crown molding, or chair rail is painted with the same flat wall paint as the rest of the wall, little or no additional painting time will be required. You can ignore that molding when estimating the wall. But when molding is

Introduction to Estimating

painted separately from the adjacent wall, additional time will be required. See Figure 5-4.

Cutting-in and painting a strip of molding that runs the length of all four walls in a 12' by 15' room shouldn't take more than thirty minutes.

Typical molding
Figure 5-4

Rounded-off, this means a journeyman will paint about two linear feet of molding per minute.

Ceilings and walls— When you're working in someone's house, furnishings and other objects usually obstruct at least some of the work. This makes estimates for ceiling and walls more difficult. Assuming a minimum of obstructions and otherwise good working conditions, painting the walls and ceiling of a 10' x 12' x 8'-high or 12' x 15' x 8'-high room will take about two manhours. This figure assumes use of a roller, a minimum of brushwork, and ceiling and walls of the same color.

Living rooms and family rooms are a bit larger, usually closer to 20' x 14' x 8' high. These, too, can be painted in about three manhours.

Painting the walls a different color than the ceiling will take more time, about 25% more. For example, if you figure two hours for painting a room when the ceiling and walls are the same color, add 25% (30 minutes) when two colors are needed.

Spray painting— All the ceilings, walls, doors and cabinets in a typical three bedroom house could be spray painted in a day, once you're set up and if you're using the same color throughout. But spray painting residential interiors is usually a bad idea. It's just too messy. The exception would be an unoccupied home in which carpeting or flooring is to be installed *following* your paint job.

Cabinets— Assume a home has about 30 linear feet of kitchen cabinets — 15 feet of base and 15 feet of wall cabinet. See Figure 5-5. Painting these cabinets with one coat, both inside and out, should take four manhours. This works out to 7½ linear

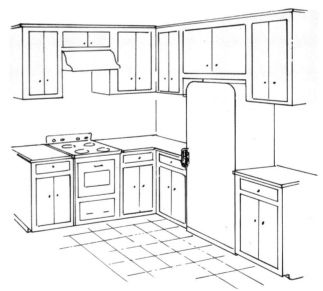

Typical kitchen cabinets
Figure 5-5

feet of cabinet per manhour. Measure the linear feet along the face of the cabinets. Allow additional time for hardware removal and whatever surface prep is needed.

Most bathrooms have five or six linear feet of base cabinet. Painting one coat should take about one manhour, again excluding surface preparation and hardware removal.

If a cabinet runs from floor to ceiling, double the linear footage. It takes nearly twice as long to paint a broom cabinet that's 3' long and 8' high as it does to paint a base cabinet that's 3' long and 36" high. If you paint 8'-high cabinets for the same price as 3'-high cabinets, you're painting over half the cabinet at no charge. For the 3'-wide broom closet, double the linear footage to six feet and base your price on that figure.

Estimating Stainwork
Staining is estimated about the same way as other painting jobs. You have to decide how much time will be required to do the job, how much the materials will cost and how much overhead and profit to add.

But there's a major difference when estimating stainwork. The possible range in quality and craftsmanship is much greater than in other painting. Many people are willing to pay premium prices to get premium work when wood is to be stained.

Here's an example. You've received a request for a bid from an owner who's installing expensive oak cabinets in his kitchen. Normally you might charge $150 to $200 to paint these cabinets. But your client wants a top-quality stain and lacquer finish. He's willing to pay $500 for this work.

Why the huge price difference? There are several reasons. First, oak cabinets are expensive. They'll make this kitchen a showplace. Painting these decorative oak cabinets would be a crime. The extra money spent on staining is small compared to the investment already made in the cabinets.

Second, your client doesn't want to entrust the work to an incompetent who might permanently damage the wood. Top-grade professional staining takes more skill than many painters have. It takes practice, patience and a good eye for color. Doing stainwork is the Ph.D. of painting. With stainwork, the craft of painting becomes an art.

Of course, it doesn't take a magician to stain raw wood cabinets, doors, paneling, or trim. Matching stain on an antique chair, for example, is much more difficult work. But staining cabinets isn't the type of thing you'd leave to a green apprentice, either.

If you know how to do high-quality stainwork terrific. Charge more.

The type of stainwork I'm referring to is a top-quality, custom, one color finish. It's important to distinguish this from other types of stainwork that are more complex, highly technical or require specific training and skills. This might be called super-custom staining. You won't see much, or maybe any, of this type of work. This is usually done by specialized furniture finishers.

Let me suggest some prices for top-grade, one-color stain jobs with a high-quality smooth finish of lacquer, urethane, or varnish.

The basic rule of thumb is that you should charge twice as much for stainwork as for general painting. If you make $100 a day painting, you should make $200 a day staining and finishing.

I generally price cabinet staining by the linear foot. For example, in 1984 in Los Angeles, most stain and lacquer jobs went for about $15.00 a linear foot. This price includes all materials, labor, one coat of stain, two coats of sanding sealer and two coats of lacquer inside and out. Also included in this price might be an extra touch-up with stain on any areas that came out too light. For a customer who's determined to have everything "just perfect," the price could go as high as $20.00 a linear foot.

Applying varnish with a brush instead of spraying on lacquer takes more time and effort. But it's hard to pass that extra cost on to the customer, because other contractors will undercut your price. So if you choose to brush on varnish instead of spraying lacquer, be prepared to earn less money on the job. There is an exception to this. And it hinges on your ability as a salesman.

Varnish or urethane, when properly applied, is considered the "Cadillac" of finishes. It produces a warmer, more natural look than lacquer, and will not break down from exposure to water as common lacquer will. This is a strong selling point for kitchen and bathroom cabinets. Make up some nice samples (small cabinet doors are a good choice) and use them as selling tools. You should be able to charge 25 to 50% more, and get more cabinet jobs.

In Los Angeles in 1984, doors with six inset panels were going for $25.00 or $30.00 per side, including the door jamb. An eight-lite sash window would cost about the same as a door, from $25.00 to $30.00 per side. Staining a three-foot high bookshelf would go for about $7.50 per linear foot. The cost would double for a floor-to-ceiling bookshelf. For a bookcase six feet high and 12 feet long, you would multiply 12 feet by $15.00 to get a price of $180.00. If the bookcase sat on a cabinet that also had to be stained, the price would be $15 per linear foot for the cabinet and $7.50 per linear foot for the bookcase.

Prices will, of course, vary with the cost of labor and materials and competitive conditions. Until you're familiar with competitive prices in your area, base your estimates on the cost of materials and the time required to do the work. But always remember, if you have the knowledge to do stainwork, charge at least as twice as much as for other painting.

Estimating Forms

To this point, I've explained the estimating procedure, offered checklists you can use when estimating the job, suggested how to figure the quantity of paint required, and given you some recommendations on estimating manhour requirements. Now it's time to start putting dollars and cents down on paper.

Every paint estimator needs a good estimate take-off form. Figure 5-6 is the one I use. Make any improvements or alterations on it, then get a printer to run off 500 or 1000 copies with your company name on them. The printer can make them up into pads so they're easier to use. Keep a good supply on hand.

A good estimate take-off form will organize your estimate, help eliminate errors and reduce the time needed to compile the estimate. Since you're not charging for estimates, reducing wasted time is important.

To start the take-off, go to your interiors checklist. Take the first item on the checklist. Write the description of that item in the top left box of the take-off form. Fill in the color and number of coats. Then enter the time needed for setup, preparation, painting and cleanup. Total these times in the next column. Multiply your labor cost per hour by the time needed and enter the result in the column headed "Cost of Labor." Next, enter the material cost. Finally, total the labor and material costs in the last column. Figure 5-7 shows how the form might look after the first line is complete.

Fill in the form line by line from left to right. Don't skip any columns. List every work item — walls, ceiling, doors, windows, trim. I like to use one page for each room. That usually leaves several blank lines on each page, so there's room for changes if necessary.

When each page of the estimate is complete, add up the "Total Price" column. Enter the total at the bottom of the page and circle it so it's easy to spot.

At the end of this chapter there's a complete sample estimate for a new home. Notice how this form was used to compile that estimate. Any time you make a complete labor and material take-off, use the estimate form. It's your basic estimating document.

The completed estimate is your record of how the costs were developed. But it can also be your bid. There's no reason why it can't be sent to your client. It has all the information the client needs to evaluate your price. Don't waste time filling out a separate bid form. Simply make a copy of your estimate and mail it to the client.

Of course, included with the estimate that goes to your client is a cover letter that becomes the contract. Later in this chapter there's a sample cover letter. The cover letter proposes to do the work listed on the estimate and at the price stated in that estimate. Your estimate then is *incorporated* into the contract. It becomes part of the agreement between you and your client.

Here's why you want that to happen. Your estimate defines the scope of the job. Anything you've omitted from the estimate isn't included in your price. You promise to do just what's on the estimate, nothing more. If you've missed something important, it's your client's responsibility to find it. If both you and your client discover later that something was left out of the estimate, that becomes extra work which increases the amount due on the job. That's an important protection that every painting contractor should have.

Common Materials Estimate

Your bid price includes more than just the labor and material costs on the estimate form. There are other costs involved in every job. And, as you'll remember, we're going to identify every cost for this job and put a number down by it.

A common estimating mistake is to omit (either intentionally or accidentally) the expendables like masking tape, sandpaper, spackle, rags, and thinner. But these costs add up fast, especially on larger jobs. Don't forget to charge for them.

Paint Contractor's Manual

Residential Estimate Form

Item	Color	Number of coats	Time for setup	Time for prep	Time to paint	Time for cleanup	Total time	Cost of labor	Cost of material	Total price

Estimate take-off form
Figure 5-6

Residential Estimate Form

Item	Color	Number of coats	Time for setup	Time for prep	Time to paint	Time for cleanup	Total time	Cost of labor	Cost of material	Total price
4 WINDOWS 12 LITES EACH	WHITE	2	20 MIN.	15 MIN. EACH = 1 HOUR	30 MIN. PER COAT 2 COATS PER WINDOW = 4 HOURS	10 MIN. EACH = 40 MIN.	6 HRS.	120.00	1 QUART OF WHITE ENAMEL = 4.00	124.00

Estimate take-off form in use
Figure 5-7

Some painting contractors include items titled "expendables" and "small tools" on every estimate. These expendable items are carried over from job to job. The cost of replacing brushes, drop cloths, spackle blades and rollers would fall under small tools. A 2% or 3% allowance will cover expendables and small tools on most jobs.

I prefer to identify a dollar cost for small tools and expendables in my estimates. This takes a little more time, but it's relatively easy if you make up a form like Figure 5-8. The materials are preprinted on the form, but the type, quantity and cost columns are blank. I fill in those columns when I make the estimate. There are blank lines so you can add other common items that apply to your jobs. You'll find a blank copy of this form, and the other customized forms used here, in the back of the book. Adapt, or adopt, them for your business, and have copies printed, as described earlier.

The list does double duty. First, it keeps the cost of most commonly-used materials at hand. I plug these costs into the estimate as needed. Then, when the estimate is complete, the form becomes a shopping list of what's needed for the job. When I go to the paint store to pick up materials, this form is my shopping list. I check off each item as it's loaded into my truck. That cuts the number of trips back to the paint store for items I forgot on the first trip.

Paint Shopping List

The common materials estimate doesn't include paint needed to do the job. That's on the paint shopping list, Figure 5-9. Develop this list from the figures on the estimate form, Figure 5-6. Figure 5-9 is just what the title says, the paint shopping list for use at the paint store. Note that there's no need to put prices on this list. They're already noted on the estimate form. The shopping list will help you get the right paint to the right job on the right day.

Equipment Estimate

Use a form like Figure 5-10 to list tools and equipment that you will need to rent or buy for the job.

Estimate Summary

Figure 5-11 shows the estimate summary. Here's where you bring all the costs together and total them to get the bid price. The top line shows the total price taken from the estimate form, Figure 5-6. The next two lines show the totals from the common materials and equipment estimates. Then there's a line to subtotal these figures so you can compute your overhead and profit. Enter your overhead and profit figure, then total the "Price" column to get the bid price.

Overhead

Overhead and profit appear as a single line on your estimate summary. That's an important line. Let's look at overhead first.

Overhead expense is a major part of every job. Overhead is the cost of staying in business: office rent, phone, advertising, office help, heat and light, office insurance, cars and trucks, stationery, postage, accounting and legal fees, non-productive labor (supervision), interest, repairs and maintenance, bad debts, depreciation, your time spent compiling estimates, the cost of this book, and countless other items. Just about every expense that isn't the direct result of taking on some particular job is part of your overhead. Add these

Common Materials Estimate

Item	Type	Quantity	Cost
Sandpaper	220	1 SHEAF	$17.00
Sandpaper	50	10 SHEETS	4.00
Spackle	INT.	1 QUART	3.00
Patching material		1 GALLON	5.00
Putty			
Caulk			
Paint thinner		5 GALLONS	10.00
Lacquer thinner			
Rags		3 POUNDS	6.00
Roller	3/4" NAP	3	8.00
Roller	1/4" NAP	2	4.00
Tape	1"	1 ROLL	2.00
Plastic drops			
		TOTAL	$59.00

Common materials estimate
Figure 5-8

costs up each month and you've got a substantial expense, even if you do business out of your truck and have no office staff.

If you're like most painting contractors, overhead will probably be about 15% of gross income. Fail to include a realistic charge for overhead in each estimate and you're giving away the store — and probably losing money on every job.

There are many ways to calculate overhead expense and include it in each estimate. Probably the simplest is to estimate your total overhead expense for the year. Then divide that figure by estimated sales for the year. The result is your overhead as a percentage of sales — the amount to add to each bid to cover overhead.

Let's say you're running a low-overhead operation. There's no rent because you do business out of your home. The only office expenses you have are a part-time bookkeeper, a phone, business insurance, a pickup truck, and some miscellaneous items. The total of these expenses will come to about $5,000 over 12 months. Another $5,000 will go for advertising during the next year. Most of the time you work along with your painting crew and draw a wage as a painter. So supervision expense is small. Still, about one-third of your time is spent estimating, managing, or selling. You figure that overhead time is worth about $10,000 a year to your company.

Paint Shopping List

Brand	Name	Code Number	Flat	Enamel	Quantity
DUNN-EDWARDS	PEARL WHITE	70	X		5 GALLONS
DUNN-EDWARDS	PEARL WHITE	70		X	1 GALLON
DUNN-EDWARDS	SUNSET BEIGE	64	X		2 GALLONS
DUNN EDWARDS	MOON LILY	05-20D		X	2 QUARTS
AMERITONE	FROSTY MORN	181H	X		1 GALLON
AMERITONE	EGGSHELL	194C		X	2 QUARTS

Paint shopping list
Figure 5-9

Equipment Estimate

Item	Description	Cost
SPRAY GUN - RENT	AIRLESS - 100' OF HOSE	$50.00
LADDER - RENT	40' EXTENSION	15.00
SCAFFOLDING - RENT	1 ROLLING STAGE - 2 LEVELS HIGH	40.00
	TOTAL COST	$105.00

Equipment estimate
Figure 5-10

Estimate Summary

Item	Price
Painting and materials	$1,500.00
Common materials	59.00
Equipment	105.00
Subtotal	1,664.00
Overhead and profit - 25 %	416.00
Total	$2,080.00

Estimate summary
Figure 5-11

Introduction to Estimating

Adding office costs ($5,000), advertising ($5,000) and your salary as the manager of your business ($10,000), you have $20,000 for overhead in the next year. If you estimate gross income to be $200,000 for the coming year, you're doing pretty well. Overhead is only 10% of sales. But if income falls to $100,000 in 12 months and overhead remains unchanged, overhead has ballooned to 20% of gross. That's probably too high.

Whether your overhead is high or low, include it in each estimate. If you discover that overhead is running 15% of gross, include that 15% in every bid. Remember our goal? *Identify every cost and include in each estimate all the costs you identify.* Overhead is a real cost of painting. There's no way to do business without overhead. That's why every estimate has to include a realistic estimate of overhead expense.

Profit

There's no way to calculate exactly how much to add for job profit. Naturally, every painting contractor wants to make as much profit as the traffic will bear. But the competition will take too much work away if your profit percentage is too high.

What's a good figure to add for profit? Some contractors insist that 10% is too low. It probably is if that "profit" is really just a way to cover all the overhead expense every painting contractor has. What I'm talking about here is *real* profit, money left over after all bills are paid and after you've paid yourself a reasonable wage.

Some contractors take work at no profit just to keep their crews busy and the cash coming in. I wouldn't recommend that.

Here's my thinking. If the estimate really covers every expense, a 5 to 8% profit is enough. Think back to our discussion of goals in Chapter 1. We decided that a 5% profit after tax was a good goal. That's all you need to build a substantial painting business over your lifetime. Include a 5 to 8% profit in each estimate and leave that money in your business. When work is plentiful, increase the profit percentage a point or two. Cut back a little if you're losing too many of the jobs you would like to have.

Estimating Packet

If you don't already have good estimating forms, use the checklists and forms in this book. My suggestion is to bind the forms together into an "estimating packet." The binder for your packet could be a three-ring loose-leaf notebook with pouches for holding paper on the inside cover.

Make copies of each of the checklists and forms. Punch holes in the forms so they fit in a three-ring binder. Put the forms and checklists in the notebook and separate them with dividers. Label the dividers for quick reference.

Put your reference materials, the estimating checklists, estimating tips, and painting time list in the front of the notebook. Keep them separate. Next would come the estimate forms, then the common materials and equipment estimates, estimate summary, and, finally, the paint shopping list.

Use the pouches on the inside cover to hold blank note paper, pens and a calculator.

Fill out the forms and checklists as you make the estimate. When the estimate is complete, take it out of the notebook and put it in a manila folder. File it in your office file of "Estimates Pending" with the customer's name written on the tab.

Making the Bid

Now that your estimate is complete, it's time to submit the bid. It's best if the cover letter for your bid is typed on your company stationery. If you don't own a typewriter or don't type, print the bid neatly on letterhead. A scrawled, barely readable bid makes a poor impression and will probably be discarded.

Here's a trick to help you print neatly. Put a piece of lined paper under your stationery when writing up the bid. You can see the lines through the top sheet. Print along these lines so the handwriting stays straight.

It's extremely important to explain exactly what the bid covers and what the bid excludes. A good job of describing what you plan to do will eliminate about 90% of all disputes and will help you get paid sooner at least 10% of the time. In business, you don't rely on memory and verbal agreements. Put everything in writing and you'll avoid disputes and misunderstandings.

Until signed by you and your client, the bid isn't a contract. It does, however, become the contract once accepted. That's why the bid should be written, and should cover all details.

Figure 5-12 shows what the cover letter for your bid might look like.

ALLWAYS PAINTING COMPANY

October 15, 1984

Mr. John Doe
5700 3rd Street
Jump Off, Arizona 00101

Dear Mr. Doe:

Our bid price for painting the interior of your home at 5700 3rd Street is $2,080.00. This is our complete price, including all materials and labor as itemized on our attached estimate.

We will prepare and paint all surfaces in a manner that meets professional standards. This price covers the following work:

Living room:
    Ceiling and walls -- 2 coats flat Navajo White
    Two doors and jambs -- 1 coat enamel Navajo White
    Four windows -- 1 coat enamel Navajo White

Dining room:
    Ceiling and walls -- 2 coats flat Navajo White
    One door and jamb -- 1 coat enamel Navajo White
    Chair railing -- 1 coat enamel Navajo White
    Crown molding -- 1 coat enamel Navajo White
    Two windows -- 1 coat enamel Navajo White

Kitchen:
    Ceiling -- 2 coats enamel Dunn-Edwards Lemon Mist/31
    Walls -- 1 coat oil-base primer for wallpaper
    Two doors and jambs -- 2 coats enamel Dunn-Edwards Lemon Mist/31
    One window -- 2 coats enamel Dunn-Edwards Lemon Mist/31
    Interior and exterior of cabinets -- 2 coats enamel Dunn-Edwards Lemon Mist/31

These prices are for acceptance within 10 days. They assume that furniture and personal belongings will be moved at least three feet away from all surfaces to be painted before our crew arrives. Please remove all hardware, latches and handles before work begins. We will spend two or three days painting and will start within 10 days of your acceptance of this proposal. The full contract price is due upon completion of the work.

To accept this proposal, just date and sign one copy of this letter and return it to us in the envelope provided.

Thank you for giving us the opportunity to bid on this work.

Sincerely,

Ralph Smith

Accepted by: _____
Date: _____

11458 BURBANK HIGHWAY
PHONE 271-2300
LICENSE #000-321

JUMP OFF, ARIZONA
ZIP 00101
BONDED/INSURED

Bid cover letter
Figure 5-12

Using Standard Contracts

Printed contract forms for painting or construction are available at larger stationery stores. These forms include a lot of fine print that might be helpful to you if a dispute develops. But they never allow enough room for writing down all the work to be done.

The trick is always to describe in detail exactly what's included in the bid. If your contract just gives a price for painting "five rooms," there's going to be a dispute about what's included and excluded and about colors that don't match. Instead, describe colors by brand and name or number and specify what surfaces get painted — which walls, what trim, which windows and doors, etc. If it isn't on your list or included in the estimate you submit with the proposal, it isn't going to get painted. There shouldn't be any misunderstanding. If your client later discovers that you forgot to include a handrail in the bid, that's O.K. You can paint that too. But it will be an extra cost.

In other words, make the contract include *only* what you say it includes. Anything you forgot is excluded from the bid, and won't be painted by you at no charge.

As I suggested, the best way to limit the scope of the job is to make your estimate a part of the bid. The letter you write to the client describes the work to be done "as itemized in our attached estimate." Your client gets a copy of the estimate and knows exactly what your bid covers. If you feel that there's a good chance of a dispute, ask him to initial each page of the estimate to indicate his acceptance.

Your client's signature on the bid makes it a legal contract. Both you and your client should keep a copy to resolve questions about what work is to be done.

Payment Terms

Every contract should specify a time when payment is due. Be sure to cover this important subject. On a larger job, get a portion of the bid price in advance. Your schedule might look like this: 10% upon signing the agreement, 50% upon completion of exterior flat work, 30% upon completion of exterior windows, doors, wood trim and gutters, 10% upon completion of all touch-up work.

If you use a standard contract, it will probably call for final payment 10 days after completion of the job. If so, cross out the words, "within 10 days of" and write in "promptly upon." It should read, "Final payment to be made promptly upon completion of the job." You won't always get paid on completion, but without this statement in the contract, payment could come much later.

Homeowners and smaller contractors can usually be persuaded to pay promptly about 90% of the time. Sometimes even the big companies will agree to it.

Don't leave collecting to chance or the customer's good will. Never encourage payment by mail. In ten years of painting, I've had only one check arrive on time through the mail. Collect your money in person.

The contract is a legally binding agreement between you and the customer. Be sure it covers everything agreed on by you and your customer. Don't have side agreements or verbal understandings that supplement or replace parts of the written agreement. Put every detail in writing: all the work you're going to do, the money you're to be paid, and when the payments are due.

Once you've suffered through the upsets, headaches, frustrations, and financial losses of a job done without a written contract, you'll appreciate the advantage of a well-written contract.

Estimating Apartments

Estimate the interior and exterior of apartment buildings the same way as other residential property. The price range, however, differs because the quality of work is generally lower. Unless you're dealing with very classy rental property, landlords don't want to pay as much to paint their apartment as they would their home. You will find exceptions to this, but they're in the minority. Generally, the price for rental property is about half of what you would charge for a residence.

To make money in apartment painting, the job has to be done about twice as fast as a good-quality residential job. As you might expect, quality may suffer. But you must still maintain an acceptable standard of neatness and get full coverage on all painted surfaces.

Repainting apartments is considered by some to be the slop art of the painting profession. I know guys that can paint two one-bedroom apartments in a day. That's definitely high-speed work.

Larger Commercial Buildings

Large buildings are estimated the same way as

small homes. Figure each area to be painted and then add all the parts together to find the total price. On a larger building, it's absolutely essential to use this method to avoid total confusion.

Let's say you have to estimate a thirty-two unit stucco apartment building that is three stories high. Each unit has a balcony.

Start with the exterior. Find the square footage by multiplying the length of each wall by the height. Figure 5-13 shows how to compute surface area. Be sure to include areas inside balconies and any passageway overhangs.

Next, compute the time (in manhours) needed for prep work on the stucco. To do this you have to know how fast your crew can work, what standard of quality this particular job requires and exactly how you're going to do the prep work.

Figure high productivity on this job. Apartment painting is very competitive. It requires a fast, hard-working crew and a careful estimate to make any money.

You've checked the stucco for the preparation needed. You figure that it needs sixty manhours of prep work by a competent journeyman. Multiply sixty hours by your journeyman rate to get the labor costs for the prep work.

Let's say your labor cost per journeyman manhour is $20, including the base wage, fringe benefits, all taxes and insurance. Multiplying $20 by 60 gives us $1,200.00. Add in your material cost for surface preparation, overhead and supervision cost and you've got an estimate for the stucco prep.

Next, we'll estimate the stucco painting cost. Assume that the total stucco area is 100,000 square feet. A gallon of paint will cover 300 square feet of stucco. Divide 100,000 by 300 and you come up with 333.3 gallons needed to do the job. To find the manhours needed for stucco painting, first decide how the painting will be done. Will it be sprayed, brushed or rolled? Will you work off ladders and picks or rolling scaffolding? Once you've decided this, figure how long it will take to paint each area of the building. Be sure to include time spent covering adjacent surfaces with drop cloths and getting tenants to remove their plants and furniture from the balconies.

Suppose you figure that five journeymen can paint all the stucco in a forty-hour week. That's 200 manhours. Multiply 200 by $20 per hour to get a labor cost of $4,000.00. Then add in the cost of materials, supplies, any equipment rental, supervision, overhead and profit to get the total price for stucco painting.

Complete the rest of the estimate using the same procedure. Break each step down into its labor and material component. Multiply the units painted by the labor or material cost per unit, add the cost of supplies, equipment rental and supervision, if any, and then add in your overhead cost. Put all this information down line by line on your estimate form. Refer to the estimate checklist to make sure you've covered everything.

Keep this in mind whenever you estimate a large building. Nine times out of ten, the owner will take the lowest bid. That's O.K. You can do very well painting exteriors for low prices if you're fast and figure the job that way. You'll have to plan on higher production than when estimating single homes.

With some experience, you'll learn how to price building exteriors competitively. To speed up this process, always find out what the other bids were — after the job has been awarded, of course.

Reading Blueprints to Estimate New Construction
You'll have to read blueprints if you want to estimate new construction and larger remodeling jobs. Reading construction plans isn't hard. It just requires some practice, patience and attention to detail.

Blueprints are drawn in scale, the most common being *quarter-inch scale*. That means that each 1/4" on the blueprint equals 1 foot in the actual building. If you don't already have one, get a triangular architect's rule or a tape with 1/4" scale on one side and 1/8" scale on the other. Most stationery stores and drafting supply houses offer scales or tapes like these.

Watching for Changes
Be alert for any changes in the plans. Changes are usually very important if you're a painting contractor. They usually include something like an extra set of cabinets or some extra windows. That's important information for your estimate. These changes may not be clear on the prints. Figure 5-14 shows what you're likely to get.

Introduction to Estimating

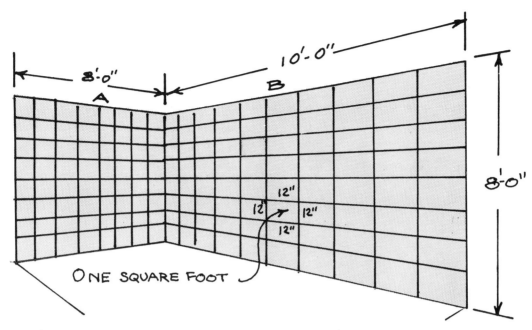

HEIGHT × LENGTH (OR WIDTH × LENGTH) = TOTAL SQUARE FEET

WALL A = 8' HIGH × 8' LONG: 8' × 8' = 64 SQ. FT.
WALL B = 8' HIGH × 10' LONG: 8' × 10' = 80 SQ. FT.

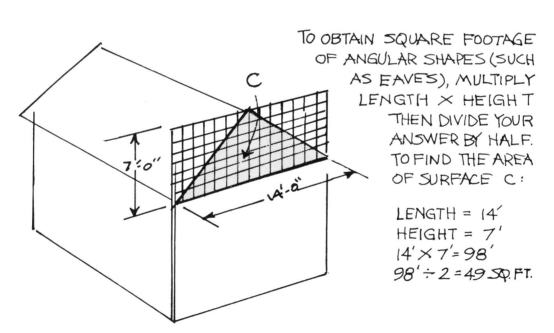

TO OBTAIN SQUARE FOOTAGE OF ANGULAR SHAPES (SUCH AS EAVES), MULTIPLY LENGTH × HEIGHT THEN DIVIDE YOUR ANSWER BY HALF. TO FIND THE AREA OF SURFACE C:

LENGTH = 14'
HEIGHT = 7'
14' × 7' = 98'
98' ÷ 2 = 49 SQ. FT.

Computing surface area
Figure 5-13

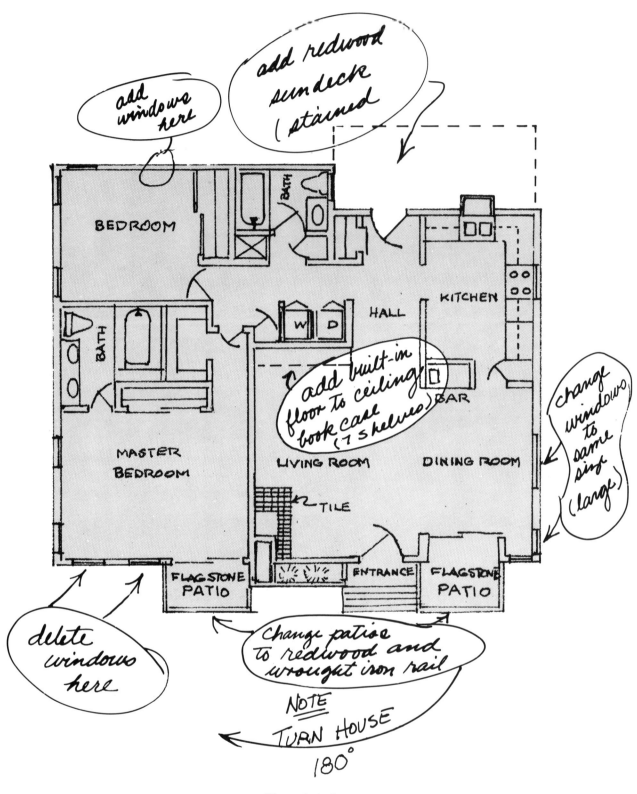

Blueprint changes
Figure 5-14

Introduction to Estimating

The contractor will probably ask for your questions. But just in case he forgets, make a note to question him when you've finished looking over the prints. Write down any questions that come up as you compile the estimate. Make it your policy to ask the general contractor, "Is there anything that's not shown on the prints?" This is especially true of remodeling work and custom homes where a homeowner is involved. Homeowners who double as amateur architects have been known to change the plans several times in a single day. To protect yourself from blueprint changes or additions, always include in your estimate and contract the date and copy number of the blueprints you bid. Any changes or additions made after that date become subject to extra charge.

Blueprints can look terribly complicated. They may be for the other trades. But fortunately, as a painting contractor, your job is relatively simple. You're concerned with surfaces only and these are usually well-defined on the plans.

The worst thing you can do when going over a set of plans is to rush through them haphazardly. It's easy to overlook a set of cabinets here or a

---

Blueprint Take-off Checklist

1) Measure the floor plan for total square footage of living space. The scale of the drawing (for example, ¼" = 1') will be noted in the lower right hand corner of the page.

2) Count the number of doors on the floor plan. Then refer to the detail section and finish schedule for data such as the number of each type of door, door style and dimension.

3) Count the windows the same way you counted the doors.

4) Check the floor plan for cabinets. Check the kitchen, service porch, and bathrooms. Look in the hall or bath for linen cabinets. Check the bedroom, living room and den for bookcases. Turn to the cabinets detail section for complete data on dimensions, specifications, and materials of construction.

5) Search the floor plan for detail items such as decorative trim, paneling and beams. If you don't find any, check the details section anyway. Sometimes they'll be missing from the floor plan but will show up in the details section.

6) Look over the exterior elevations section to determine what exterior painting will be required. Check for fascia board, overhangs, wood siding, air vents, window trim, sheet metal caps, and pipes and vents on the roof which will require painting. Also check for such exterior items as fences, balcony railings, wrought iron and patio trestles, or lattice work.

7) Check the finish schedule and specification to find out what types of paints and primers are to be used and how many coats of each are to be applied to the specified surfaces.

8) Take your list of written questions to the general contractor or architect to get answers. Don't stop asking questions until all your questions are answered. As a precautionary measure, make the last question, "Is there anything else about the paint job that isn't shown on the prints?"

9) Be sure you have the list of colors to be applied. This is rarely listed in the prints, so you'll have to ask for it.

Blueprint take-off checklist
Figure 5-15

Paint Contractor's Manual

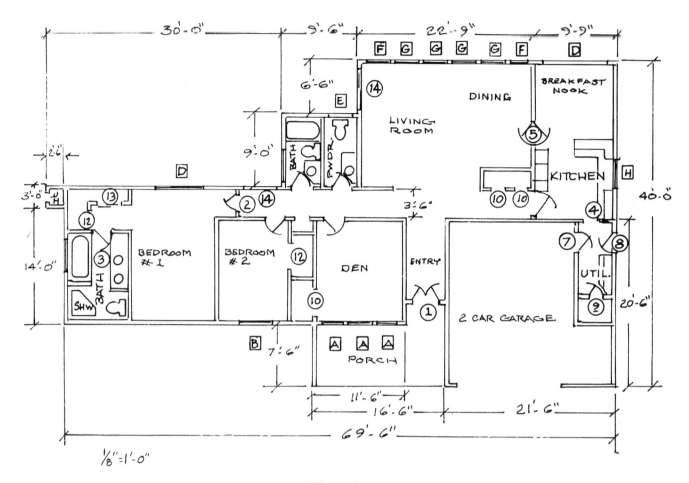

**Floor plan
Figure 5-16**

beamed ceiling there. Those are costly mistakes. So take your time. Be thorough. Don't hesitate to ask questions if you're confused or not totally certain about something on the plans.

As we do our take-off, we'll use the blueprint take-off checklist, Figure 5-15.

Sections

Painting contractors can ignore some of the plan sheets. For example, you don't learn much about the painting required by studying the excavation plan or foundation plan. But other plan sheets are essential. Most of what you need to know will be on plan sheets labeled "A" in the lower right corner. These are the architectural sheets that show the building itself. If you don't find what you need on these sheets, in the details, on the finish schedule, or in the specifications, the data is probably missing. Ask the contractor or the architect for clarification.

The paragraphs that follow describe the various plan views you'll use when making a take-off.

The *floor plans* show the length of all walls. That lets you calculate the floor area in all rooms, including the garage, patio and balcony. The floor plans also show all lofts and roof decks, important items for the painting estimate.

The floor plan is a cutaway, bird's-eye view of what the interior of the finished house will look like. See Figure 5-16. It shows walls, doors, windows, cabinets, ceiling beams, paneling, special molding, staircases, stair railings, closets and whatever else might be in the house. It shows length and width but not height.

Introduction to Estimating

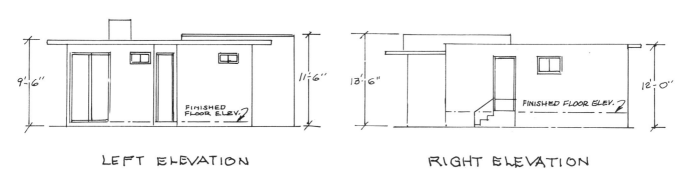

Elevations
Figure 5-17

The *elevations* show the exterior of the structure from four different viewpoints: north, south, east and west. They show all the exterior features, such as wood siding, decorative trim, wrought iron, overhangs and wood fences. The elevations show length and height of each wall and label surface materials such as stucco, redwood, or brick. See Figure 5-17.

Section views (Figure 5-18) give a cutaway view of the interior of the building as seen from the side. These drawings show the structure of stairwells and the ceiling. They give the best view of ceiling beams and the slope of the ceiling.

Details show expanded views of certain items that need special attention. Details will show things like cabinets, doors, windows, beams, paneling, stair railings, special wood trim, and overhangs. Details also show length, height, and width. See Figures 5-19 through 5-21. The *finish schedule* lists interior and exterior surfaces by type and describes what each is made of: plaster or drywall, wood or metal. Then it identifies what finish each is to receive. See Figure 5-22. Sometimes a finish schedule will list the colors to be used. But most

83

Paint Contractor's Manual

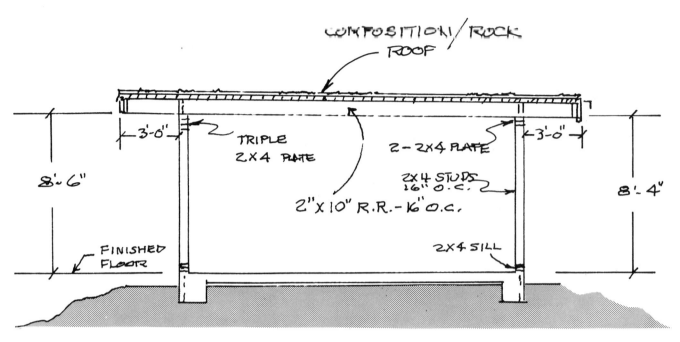

Section view
Figure 5-18

finish schedules omit this important fact. Generally, it will just specify the type of finish, not the color. For example, it might read:

    Walls----------------------------Flat paint
    Doors----------------------------Enamel
    Cabinets-------------------------Stain & Lacquer
    Baseboards-----------------------Vinyl Cove Molding
    Bathroom Walls-------------------Tile

Cabinets
Sometimes it's hard to see cabinets on the floor plan. They may be represented by a rectangle with a diagonal line drawn between two opposite corners. But you'll see a frontal view in the cabinet details. Here you'll find the kitchen cabinets, bathroom pullmans, bookcases, and all other cabinetry in the project.

The cabinet detail section shows height, width and length of the cabinets, plus some of the ornamental details. This is important. It's the ornamentation that makes a cabinet hard to prep and paint.

Paneling and Decorative Trim
Decorative trim such as crown molding or wainscoting won't be on the floor plan. You have to turn to the detail section to find it. There will be a drawing of the wood trim or crown molding, for example, but it will show only a section view, just enough to hint at the face view. Beside the section view will be the dimensions and the type of wood.

If there's any paneling or wood trim wall panels, they will be shown more completely in the details. This tells the contractor exactly how much there is and where it's located. In some cases, entire rooms will be detailed. On other projects, just two walls or perhaps only a fireplace mantle will be drawn. This may be all that's needed to show the details.

In the detail drawings you'll also find the dimensions of the trim or paneling and a description of the wood to be used.

Beams and Wood Ceilings
Beams and wood ceilings appear on the floor plan as a series of lighter-colored broken lines which are

Introduction to Estimating

drawn across the ceiling area. Usually there will be a symbol such as "B-1" next to the broken lines. That means that detail drawing B-1 depicts those beams. An elaborate beam or all-wood ceiling may be shown in full so there's no doubt about its construction and design.

Most floor plans are crowded with symbols and notes in addition to the lines and numbers that show components and their dimensions. The light-colored broken lines representing ceiling beams may be hard to spot in this jumble. So take your time. Look carefully for these broken lines, especially on a detailed custom house with a mumbo-jumbo set of plans.

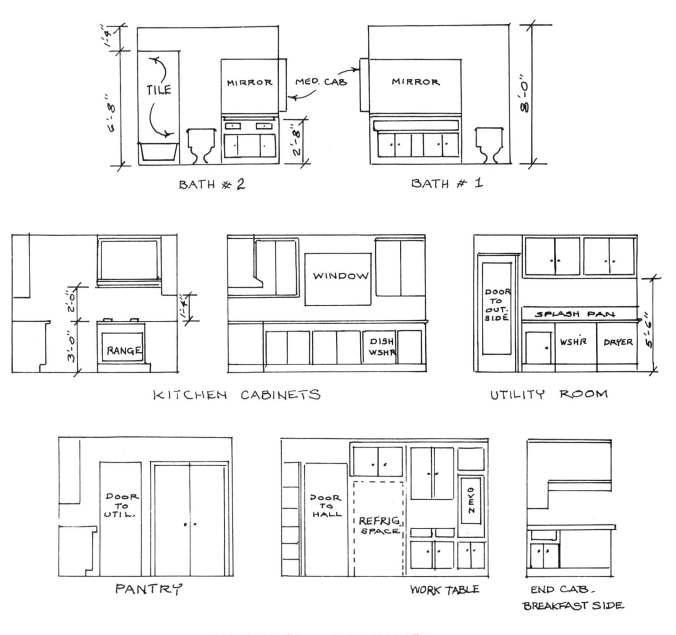

Cabinet details
Figure 5-19

Paint Contractor's Manual

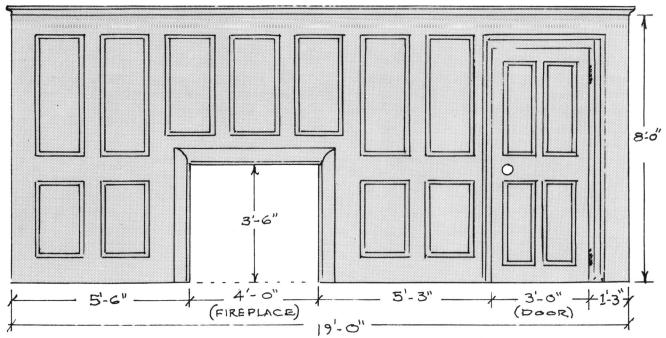

Paneling details
Figure 5-20

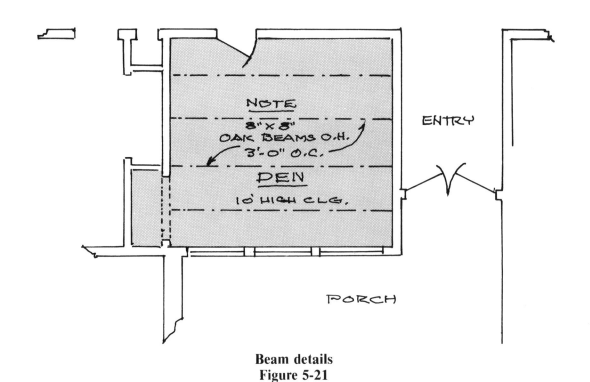

Beam details
Figure 5-21

ROOM FINISH SCHEDULE				
ROOM	FLOOR	WALLS	CEILING	REMARKS
ENTRY	VINYL TILE	PLASTER-PAINT	ACCOUS-PLAS.	
LIVING ROOM	CARPET	" "	" "	
DINING ROOM	"	" "	" "	
BEDROOMS	"	" "	" "	
HALLS	"	" "	" "	
KITCHEN	"	PLASTER-ENAM.	SMOOTH PLAS-PT.	
BREAKFAST	"	" "	" "	
UTILITY	"	" "	" "	
POWDER	"	" "	" "	
BATHROOMS	"	" "	" "	
CLOSETS	CARPET	PLAS. & PAINT	PLAST & PAINT	

Finish schedule
Figure 5-22

When doing a take-off from a set of plans, use the checklist in Figure 5-15. That makes it unlikely that you'll miss something important. And keep a blank piece of paper handy while you're doing the take-off. Write down every question you have. If the plans don't answer your questions, get the general contractor, architect or owner to provide the answer. Don't submit a bid as long as you have doubt about what's required.

Estimating Windows and Doors
Count the number of windows on the floor plan or elevation drawings. Next to each window will be a letter enclosed in a square. The letter refers you to that same window as it is listed in the window schedule. See Figure 5-23. For example, a window identified as "A" on the plans will be the window on line "A" in the window schedule.

The elevations will show the windows from an exterior view. At a glance, you can count the windows and see whether they are sash, sliding or fixed glass. But the elevation doesn't reveal whether the windows are wood, aluminum or plastic. Turn to the window schedule or specifications for this information.

WINDOW SCHEDULE

MARK	SIZE	TYPE	GLAZING	REMARKS
A	4'-0" X 10'-0"	COMB. FIX-GLASS - LOUVERS BELOW		SEE FRONT ELEVATION
B	6'-0" X 4'-0"	ALUM. SLIDING		
C	4'-0" X 2'-0"	ALUM. SLIDING		OBSCURED GLASS
D	8'-0" X 4'-0"	ALUM. SLIDING		TINTED GLASS
E	2'-0" X 3'-0"	ALUM.-LOUVRED GLASS		OBSCURED GLASS
F	4'-0" X 4'-6"	ALUM. SLIDING		TINTED GLASS
G	4'-0" X 4'-6"	FIXED GLASS		TINTED GLASS
H	4'-0" X 3'-0"	ALUM. SLIDING		
J				
K				
L				
M				
N				
P				
Q				
R				
S ETC.				

Window schedule
Figure 5-23

It's very important to note the type of window. Aluminum sash windows are common now, and even windows made of plastic are appearing. New aluminum sash rarely has to be painted. But plastic windows often do, and can be more difficult to paint than wood windows. So the type of window can make a sizeable difference in the painting cost.

Count the doors on the floor plan just like you counted the windows. Doors are represented by openings in the wall with an arc showing the direction of swing. A number enclosed in a circle accompanying each door refers you to the appropriate detail drawing and the door description on the door schedule. See Figure 5-24. The detail drawing will show the full face of the door. The type of material and door size will be indicated in the door schedule.

The window and door schedules should list all windows and doors by size, type, material and construction. Note that aluminum sash windows and some other types of windows are primed at the factory and have removable imitation sash. That's important information when preparing a painting estimate.

Note carefully the type of each wood door. Some require more work than others. Birch, for example, needs a sealer coat before being stained. That's not necessary with some hardwoods.

Door Schedule

MARK	SIZE	THICK	TYPE-MATERIAL	GLAZING	REMARKS
1	1 PR.- 6'-0" x 6'-8"	1 3/4"	HARDWOOD SLAB W/APP. MOLD.		
2	2'-8" x 6'-8"	1 3/8"	H.C. MASONITE SLAB		
3	2'-4" x 6'-8"	1 3/8"	" " "		
4	2'-6" x 6'-8"	1 3/8"	SLID'G - H.C. MASONITE SLAB		
5	2'-8" x 6'-8"	1 3/8"	DBL. ACT'G H.C. MASONITE SLAB		
6	2'-6" x 6'-8"	1 3/8"	H.C. MASONITE SLAB		
7	2'-8" x 6'-8"	1 3/8"	SOL. CORE SLAB		SELF CLOSING
8	2'-8" x 6'-8"	1 3/4"	SLAB DOOR W/LOUVER INSERT		
9	2'-0" x 6'-8"	1 3/8"	H.C. MASONITE SLAB		
10	4'-0" x 6'-8"		BI-FOLDING		
11	NONE				
12	5'-0" x 6'-8"		BI-FOLDING		
13	5'-0" x 6'-8"		SLD'G MIRRORED DOORS		
14	6'-0" x 6'-10"		2-SECT. ALUM. SLID'G		TEMPERED TINTED GLS
15					
16					
17					

Door schedule
Figure 5-24

The Specifications

The specs tell you in words (rather than by a drawing) exactly what's required. The specs list surfaces to be painted, explain how they are to be painted, what type of finish is to be used and how many coats are to be applied. Some specs identify what brand or quality of paint is to be used. Figure 5-25 is a sample set of specifications.

Specs usually include a lot of *boilerplate* that no one intends to follow to the letter — until there's a dispute. Then the specs will be enforced against the contractor or subcontractor. Unless you're doing a job for the government, some corner-cutting will be tolerated, as long as you do good quality work.

Guessing what part of the specs can be ignored with little risk requires some experience. Usually specs are not enforced to the letter on spec houses, tract houses, condos and commercial buildings. In other words, on jobs other than custom-quality painting, you have some latitude in following the specifications. Within limits, you can comply with the spirit, if not the exact letter, of the specs on many jobs.

sample painting

SECTION 09900

PAINTING

PART 1. GENERAL

Conditions of the contract will be considered a part of this section.

1.01 **DESCRIPTION**
- A. Provide all labor, materials and equipment required to complete all painting and finishing required by the Contract Documents.
- B. Related work specified elsewhere.
  1. Shop primers
  2. Factory finished materials and equipment
  3. Tile work
  4. Acoustical work

1.02 **QUALITY ASSURANCE**
- A. Comply with state and local regulations governing the use of paint materials.

1.03 **SUBMITTALS**
- A. Refer to Division 1 for procedures.
- B. Colors
  1. Color schedule will be furnished by the Architect.
  2. Architect must approve final colors furnished prior to any application.

1.04 **PRODUCT HANDLING**
- A. Deliver materials to job site in unopened containers bearing manufacturer's name and product description.
- B. Store materials in dry, clean, well ventilated area. Close containers.

1.05 **PROJECT CONDITIONS**
- A. Protection
  1. Protect floors and all adjacent surfaces from paint smears, spatters and droppings. Use dropcloths to protect floors. Cover fixtures not to be painted. Mask off areas where required.
  2. Hardware: Insure that hardware is removed before painting is started and replaced only when paint finishes are thoroughly dry.
- B. Environmental Requirements
  1. Comply with manufacturer's recommendations for environmental conditions under which coatings and coating systems can be applied.
  2. Do not apply finish in areas where dust is being generated. Apply no paint in rain, fog or mist, or when temperature is below 50 degrees F.

PART 2. PRODUCTS

2.01 **MATERIALS**
- A. Materials necessary to complete painting are listed by material number and names and are standards for kind, quality and function, taken from the stock list of architectural finishes of the Dunn-Edwards Corporation, Los Angeles, California.

Sample specifications
Figure 5-25

specifications guide

PART 3. EXECUTION

3.01 **CONDITION OF SURFACES**

A. Examine surfaces scheduled to receive paint and finishes for conditions that will adversely affect execution, permanence and quality of work. Do not apply paint or finish until conditions are satisfactory.

3.02 **PREPARATION**

A. Prepare surfaces in a skillful manner to produce finish work of first class appearance and durability.

B. Clean surfaces free of dust, dirt, oil, grease and other foreign matter prior to the application of the prime coat.

C. Repair all voids, nicks, cracks, dents, etc. with suitable patching material and finish flush to adjacent surface.

D. Primed ferrous metal. Remove all foreign matter. Prime scratched and abraded areas with BLOC-RUST Red Oxide Primer (43-4).

E. Galvanized metal. Remove all foreign matter and clean entire surface with mineral spirits. Pretreat with VINYL WASH PRETREATMENT (42-36). Apply primer the same day as pretreatment is applied.

3.03 **APPLICATION**

A. Apply material evenly, free from sags, runs, crawls, holidays or defects. Mix to proper consistency, brush out smooth, leaving minimum of brush marks, enamel and varnish uniformly flowed on.

B. Apply paint by brush, roller or spray.

C. Employ coats and undercoats for all types of finishes in strict accordance with recommendations of the paint manufacturer. All undercoats shall be tinted slightly to approximate the finish color.

D. Allow each coat to thoroughly dry before succeeding coat application. For oil paints allow at least 48 hours between coats of exterior work, except where otherwise recommended by manufacturer.

3.04 **CLEANING**

A. Remove all surplus materials and debris from site, at completion of each day's work. Remove all spatterings from all finish surfaces. Leave paint storage spaces in a clean and finished condition.

3.05 **PAINT SCHEDULE**

A. Finish surfaces in accordance with the following procedure for the surface and finish desired. Catalog names and numbers refer to products manufactured by Dunn-Edwards Corporation. Numbers used to identify paint type indicates the paint in white. Same material shall be color selected.

Sample specifications
Figure 5-25 (continued)

This isn't as terrible as it sounds. Here's an example. The specs call for two coats of flat paint over textured drywall and one coat of undercoater plus two coats of enamel on the doors. If you bid the job the way the specs require, you'll never get the work. But the owner, architect and general contractor don't really care how many coats are on the wall or doors. They just want good coverage. Other painting contractors will bid it as one coat on the walls and an undercoat and one coat of enamel on the doors.

Here's why multiple coats may not be an advantage. Manufacturers today are producing paints which will cover in one coat. If the manufacturer doesn't recommend three coats, why apply them?

In all but top quality custom work, painting specs need not be taken literally. But there's a catch. Builders will be perfectly happy to let you do the job in one coat. (Note however that enamel on wood always requires an undercoat.) But you'll always have to get "coverage," even if no specific number of coats is required. That means you are obligated to paint the walls until they are fully covered, regardless of how many coats that takes.

Since one coat coverage is possible with many off-white paints, coverage is seldom a problem. The paint will cover unless you get it too thin. This has been demonstrated thousands of times. Substituting "to coverage" for the coats required by the specifications can save some money without sacrificing quality. The contractor still protects himself from fly-by-night painters by requiring good coverage.

If you're working with a builder for the first time, make him be specific about this. You might say to him, "On the flat paint, is that to be two coats or to coverage?" His answer is your clue to what's really required.

But I wouldn't ignore any specification on a custom job. Suppose the specs on a custom job require three coats. If the builder tells you to apply only two, get that in writing over an authorized signature. If there's any squabble, you're legally bound to comply with the specs.

Whenever the general contractor tells you to just do the job "to coverage" on a custom quality job rather than actually applying two coats, verify that the architect or owner agrees with that decision.

It's rare in production-type work to require two finish coats. But it happens. If you have a high production job that requires a specific number of coats, protect yourself by asking questions.

Estimating by the Square Foot

Many contractors bid new construction by the square foot of floor area. This method was discussed earlier in the chapter. When you first start estimating from plans, it's a good idea to break the job down room by room and compile the labor and material cost for painting each part of each room. This method produces the most accurate estimates possible. But it also takes more time than pricing by the square foot.

As you develop more experience, you'll notice that estimates for similar work usually vary with the size of the house. Check this by converting your detailed labor and material cost estimates into a cost per square foot of living area. You'll find that the square foot cost is very similar on many jobs. Use these square foot costs as a check on future detailed cost estimates. Eventually, you'll be comfortable enough with square foot costs that you may rely on them entirely for some types of work.

Suppose that by using the labor and material cost method you come up with an estimate of $4,000.00 for an ordinary spec house. The living area of the house is 2,500 square feet. Divide $4,000 by 2,500 square feet and you get a figure of $1.60 per square foot. Suppose also that you do the job and make a reasonable profit. Now you have a price guide for that type of spec house. Each time you handle a job like this, refine your price so that it becomes more accurate. In time you may discover that $1.60 is a bit high when compared with what other painting contractors are charging. Adjust your estimates accordingly.

You can find the living area by multiplying the length of each room by the width and then adding the area of all rooms. This will give you the exact square footage. But it's faster to multiply the length of the entire building by the width. Then subtract out any irregularities if the building footprint isn't a square or rectangle. This is fast and only slightly inaccurate. The small error isn't large enough to cause you or the builders any concern. Ignore the garage, patio and balcony in your calculation. Include only areas which are used for living space.

I use Figures 5-26 and 5-27 as take-off forms when I plan to price the job by the square foot. As I gather information from the plans, I enter the description of each item to be painted, what coating is to be applied, and the number of coats in the appropriate box. When the take-off is complete, I have a condensed and organized list of the facts needed to buy materials and do the painting.

Figures 5-28 and 5-29 show what the interior and exterior take-off forms might look like when my take-off is complete. Notice that there are no manhours listed here. In fact, there are no dimensions, no material prices and no labor costs. None are considered because I have decided to price this job by the square foot of floor.

Sample Estimate
This chapter on estimating wouldn't be complete without a sample estimate. The estimating forms on pages 102 to 110 (Figures 5-30 through 5-34) show what a full labor and material cost estimate might look like.

Test your skill at making a take-off by trying to duplicate this estimate. Make up an estimate from the plans in Figures 5-16 to 5-19 and 5-22 to 5-24. Use manhour figures from this chapter and current labor and material costs. Add the cost of common materials, equipment, and your markup. When finished, you're well on the way to becoming an experienced paint estimator.

Interior Painting Take-off Form

Date estimate due: _____

Date estimate made: _____

Company estimate made for: _____
_____
_____

Person requesting estimate: _____

Address: _____

Phone: _____

Type of project: _____ Date of blueprints: _____

Total square footage: _____

Total price of job: _____

Interior specifications:

Surface	Type	Finish	Number of coats
Walls			
Ceilings			
Doors			
Windows			
Kitchen cabinets			
Pullmans			
Shutters			
Railings			

Interior take-off form
Figure 5-26

Interior Painting Take-off Form (Continued)

Surface	Type	Finish	Number of coats
Other cabinets			
Special trim			
Custom items			

Comments: _____

_____

_____

_____

Interior take-off form
Figure 5-26 (continued)

Exterior Painting Take-off Form

Date estimate due: _____

Date estimate made: _____

Company estimate made for: _____
_____
_____

Person requesting estimate: _____

Address: _____

Phone: _____

Type of project: _____ Date of blueprints: _____

Total square footage: _____

Total price of job: _____

Exterior specifications:

Surface	Type	Finish	Number of coats
Walls			
Overhang			
Doors			
Windows			
Garage door			
Shutters			
Decorative trim			

Exterior take-off form
Figure 5-27

Exterior Painting Take-off Form (Continued)

Surface	Type	Finish	Number of coats
Fascia			
Siding			
Gutters & downspouts			
Flashing & sheet metal			
Roof pipes			
Railings			
Fences			
Wrought iron			
Custom items			

Comments: _____

_____

_____

_____

**Exterior take-off form
Figure 5-27 (continued)**

Paint Contractor's Manual

Interior Painting Take-off Form

Date estimate due: __12-6-84__

Date estimate made: __11-15-84__

Company estimate made for: __SMITH CONSTRUCTION INC.__
__506 MAPLE ST.__
__L.A., CA.__

Person requesting estimate: __ROBERT SMITH__

Address: __SAME AS ABOVE__

Phone: __555-0015__

Type of project: __CUSTOM HOUSE/NEW CONST.__   Date of blueprints: __7-12-84__

Total square footage: __1916 SQ.FT.__

Total price of job: __$3832.00__

Interior specifications:

Surface	Type	Finish	Number of coats
Walls	SMOOTH PLASTER	FLAT AND ENAMEL	1-SEALER OR UNDERCOATER 2-FINISH
Ceilings	ACOUSTIC — — — — SMOOTH PLASTER IN BATHS AND KITCHEN — — —	NO PAINT —ENAMEL—	1-UNDERCOATER — 2-ENAMEL
Doors	FLUSH	ENAMEL	1-UNDERCOATER 2-ENAMEL
Windows	ALUM/SLIDERS	NO PAINT	
Kitchen cabinets	OAK	STAIN AND LACQUER	1-STAIN 2-SANDING SEALER 2-FINISH LACQUER
Pullmans	OAK	STAIN AND LACQUER	1-STAIN 2-SANDING SEALER 2-FINISH LACQUER
Shutters			
Railings			

Completed interior take-off form
Figure 5-28

Interior Painting Take-off Form (Continued)

Surface	Type	Finish	Number of coats
Other cabinets	STORAGE CABINETS UTILITY ROOM CABINETS	STAIN AND LACQUER	1-STAIN 2-SANDING SEALER 2-FINISH LACQUER
Special trim			
Custom items			

Comments: CLIENT EXPECTS TOP-QUALITY CUSTOM WORK

Completed interior take-off form
Figure 5-28 (continued)

Paint Contractor's Manual

Exterior Painting Take-off Form

Date estimate due: _____

Date estimate made: _____

Company estimate made for: SMITH CONST. (CONT.)
_____
_____

Person requesting estimate: _____

Address: _____

Phone: _____

Type of project: _____   Date of blueprints: __ _____

Total square footage: _____

Total price of job: _____

Exterior specifications:

Surface	Type	Finish	Number of coats
Walls	STUCCO	COLOR COTE / NO PAINT	
Overhang	STUCCO	COLOR COTE / NO PAINT	
Doors	FLUSH WITH MOLDING	ENAMEL	1- UNDER COATER 2- ENAMEL
Windows	ALUM/SLIDERS	NO PAINT	
Garage door	TEXTURED WOOD	ENAMEL	1- UNDER COATER 2- ENAMEL
Shutters			
Decorative trim			

**Completed exterior take-off form
Figure 5-29**

Exterior Painting Take-off Form (Continued)

Surface	Type	Finish	Number of coats
Fascia	WOOD	ENAMEL	1- UNDERCOATER 2- ENAMEL
Siding	TEXTURED	ENAMEL	1- UNDERCOATER 1- ENAMEL
Gutters & downspouts	METAL	ENAMEL	1- UNDERCOATER 2- ENAMEL
Flashing & sheet metal	METAL	ENAMEL	1- UNDERCOATER 2- ENAMEL
Roof pipes		NO PAINT	
Railings	METAL HANDRAIL	ENAMEL	1- PRIMER 2- ENAMEL
Fences			
Wrought iron			
Custom items			

Comments: ALL WOOD IS TO BE PRIMED BEFORE THE SCRATCH COAT FOR THE STUCCO IS APPLIED. FINISH COATS ARE TO BE APPLIED AFTER THE COLOR COTE.

Completed exterior take-off form
Figure 5-29 (continued)

Paint Contractor's Manual

Residential Estimate Form

BEDROOM #1

Item	Color	Number of coats	Time for setup	Time for prep	Time to paint	Time for cleanup	Total time	Cost of labor	Cost of material	Total price
WALLS	D&E MELON #67	1-S 2-F		½ HR.	2 HRS.		2½ HRS.	50.00	1 GAL-S 1 GAL-F = 30.00	80.00
1 DOOR	"	1-U/C 2-E		15 MIN.	45 MIN.		1 HR.	20.00		20.00
CEILING		NO PAINT/ACOUSTIC								
WINDOWS		NO PAINT/ALUM. SLIDERS								
									TOTAL	100.00

Residential Estimate Form

BATH #1

Item	Color	Number of coats	Time for setup	Time for prep	Time to paint	Time for cleanup	Total time	Cost of labor	Cost of material	Total price
WALLS & CEILING	D&E MELON #67	1-U/C 2-E		45 MIN.	1½ HRS.	15 MIN.	2½ HRS	50.00	1 GAL-U/C 1 GAL-E 25.00	75.00
1 DOOR	"	1-U/C 2-E		15 MIN.	45 MIN.		1 HOUR	20.00		20.00
WINDOW		NO PAINT/ALUM. SLIDER								
PULLMAN	D&E WALNUT STAIN	1-STAIN 2-SEALER 2-LACQUER		6 LINEAR FT. AT 10.00 PER FT				60.00		60.00
									TOTAL	155.00

Completed estimate form
Figure 5-30

Introduction to Estimating

DRESSING AREA

Residential Estimate Form

Item	Color	Number of coats	Time for setup	Time for prep	Time to paint	Time for cleanup	Total time	Cost of labor	Cost of material	Total price
WALLS & CEILING	D&E MELON #67	1-S 2-F		½ HR.	1½ HR.		2 HR.	40.00	1GAL-F 10.00	50.00
1 DOOR	"	1-U/C 2-E		15 MIN.	45 MIN.		1 HR.	20.00		20.00
SHELVES	"	1-U/C 2-E			30 MIN.		½ HR.	10.00		10.00
SHELF POLES		NO PAINT								
CABINET	D&E WALNUT STAIN	1-STAIN 2-SEALER 2-LACQUER		3 LINEAR FT. BY 6 FT. HIGH = 6 LINEAR FT. AT 10.00/FT. 60.00						60.00
									TOTAL	140.00

BEDROOM #2

Residential Estimate Form

Item	Color	Number of coats	Time for setup	Time for prep	Time to paint	Time for cleanup	Total time	Cost of labor	Cost of material	Total price
WALLS	D&E SUNTONE #11	1-S 2-F		½ HR	2 HRS.		2½ HRS	50.00	1GAL-S 2GAL-F = 30.00	80.00
3 DOORS	"	1-U/C 2-E		20 MIN	2¼ HRS.		2½ HRS.	50.00	2QTS.-E = 8.00	60.00
CLOSET SHELF	"	1-U/C 2-E			½ HR.		½ HR	10.00		10.00
SHELF POLE		NO PAINT								
CEILING		NO PAINT / ACOUSTIC								
WINDOWS		NO PAINT / ALUM. SLIDER								
									TOTAL	150.00

Completed estimate form
Figure 5-30 (continued)

Residential Estimate Form

DEN

Item	Color	Number of coats	Time for setup	Time for prep	Time to paint	Time for cleanup	Total time	Cost of labor	Cost of material	Total price
WALLS	D & E CORK #53	1-S 2-F		1/2 HR.	2 HRS		2 1/2 HRS.	50.00	1 GAL - S 2 GAL - F = 30.00	80.00
3 DOORS	"	1-U/C 2-E		20 MIN.	2 1/4 HRS.		2 1/2 HRS.	50.00	2 QTS. - E = 8.00	60.00
CLOSET SHELF	"	1-U/C 2-E			1/2 HR.		1/2 HR.	10.00		10.00
SHELF POLE		NO PAINT								
CEILING		NO PAINT / ACOUSTIC								
WINDOWS		NO PAINT / FIXED GLASS AND LOUVERS								
									TOTAL	150.00

Residential Estimate Form

BATH #2

Item	Color	Number of coats	Time for setup	Time for prep	Time to paint	Time for cleanup	Total time	Cost of labor	Cost of material	Total price
WALLS & CEILINGS	D & E PEARL WHITE #70	1-U/C 2-E		1/2 HR.	1 HR.	15 MIN.	1 HR. 45 MIN	35.00	1 GAL - U/C 1 GAL - E = 25.00	60.00
1 DOOR	"	1-U/C 2-E		15 MIN	45 MIN.		1 HR.	20.00		20.00
WINDOW		NO PAINT / ALUM. SLIDER								
PULLMAN	D & E WALNUT STAIN	1 - STAIN 2 - SEALER 2 - LACQUER		4 LINEAR FT. AT 10.00 FT.				40.00		40.00
									TOTAL	120.00

Completed estimate form
Figure 5-30 (continued)

Residential Estimate Form

POWDER

Item	Color	Number of coats	Time for setup	Time for prep	Time to paint	Time for cleanup	Total time	Cost of labor	Cost of material	Total price
CEILING & WALLS	D&E PEARL WHITE #70	1-U/C 2-E		20 MIN	45 MIN	10 MIN	1 HR. 15 MIN	25⁰⁰	USE FROM BATH 2	25⁰⁰
1 DOOR	"	1-U/C 2-E		15 MIN.	45 MIN.		1 HR.	20⁰⁰		20⁰⁰
WINDOW		NO PAINT/ ALUM. SLIDER								
PULLMAN	D&E WALNUT STAIN	1- STAIN 2- SEALER 2- LACQUER		3½ LINEAR FT AT 10⁰⁰/FT				35⁰⁰		35⁰⁰
								TOTAL		80⁰⁰

Residential Estimate Form

ENTRY & HALL

Item	Color	Number of coats	Time for setup	Time for prep	Time to paint	Time for cleanup	Total time	Cost of labor	Cost of material	Total price
WALLS	D&E PEARL WHITE #70	1-S 2-F		45 MIN	3 HRS.		3 HRS. 45 MIN.	75⁰⁰	2 GAL-S 4 GAL-F = 60⁰⁰	130⁰⁰
12 DOORS	"	1-U/C 2-E		2 HRS.	9½ HRS.		11½ HRS.	230⁰⁰	1 GAL-U/C 1 GAL-E = 25⁰⁰	255⁰⁰
2 CLOSET SHELVES	"	1-U/C 2-E			1 HR.		1 HR.	20⁰⁰		20⁰⁰
LINEN CABINET	"	1-U/C 2-E		½ HR.	1½ HR.		2 HR.	40⁰⁰		40⁰⁰
SHELF POLES		NO PAINT							TOTAL	445⁰⁰
SLIDING DOORS		NO PAINT/ ALUM. & GLASS SLIDERS								
CEILING		NO PAINT/ ACOUSTIC								

Completed estimate form
Figure 5-30 (continued)

Paint Contractor's Manual

LIVING RM. & DINING RM

Residential Estimate Form

Item	Color	Number of coats	Time for setup	Time for prep	Time to paint	Time for cleanup	Total time	Cost of labor	Cost of material	Total price
WALLS	D&E PEARL WHITE #70	1-S 2-F		½ HR	1½ HRS		2 HRS	40.00	1 GAL-S 2 GAL-F = 30.00	70.00
1 DOOR	"	1-U/C 2-E		15 MIN	45 MIN		1 HR.	20.00		20.00
WINDOWS		NO PAINT/FIXED GLASS & ALUM. SLIDERS								
CEILING		NO PAINT/ACOUSTIC								
									TOTAL	90.00

BREAKFAST & KITCHEN

Residential Estimate Form

Item	Color	Number of coats	Time for setup	Time for prep	Time to paint	Time for cleanup	Total time	Cost of labor	Cost of material	Total price
WALLS & CEILING	D&E PEARL WHITE #70	1-U/C 2-E		½ HR.	2 HRS. 15 MIN.	15 MIN.	3 HRS	60.00	1 GAL-U/C 1 GAL-E = 25.00	85.00
3 DOORS	"	1-U/C 2-E		½ HR.	2 HRS 15 MIN.		2 HRS. 45 MIN	55.00		55.00
WINDOWS		NO PAINT/ALUM SLIDERS								
CABINETS	D&E WALNUT STAIN	1- STAIN 2- SEALER 2- LACQUER		46 LINEAR FT. AT 10.00/FT.		—		460.00	LAQ/SANDING SEALER-5 GAL FINISH LAQ-5 GAL STAIN-2 GAL	460.00
									TOTAL	600.00

Completed estimate form
Figure 5-30 (continued)

Residential Estimate Form

UTILITY & STORAGE

Item	Color	Number of coats	Time for setup	Time for prep	Time to paint	Time for cleanup	Total time	Cost of labor	Cost of material	Total price
WALLS & CEILINGS	D & E PEARL WHITE #70	1-U/C 2-E		15 MIN.	1½ HRS.		1 HR. 45 MIN.	35⁰⁰	1 GAL-U/C 1 GAL-E = 25⁰⁰	60⁰⁰
3 DOORS	"	1-U/C 2-E		½ HR.	2 HRS. 15 MIN.		2 HRS. 45 MIN.	55⁰⁰		55⁰⁰
CABINETS	D & E WALNUT STAIN	1-STAIN 2-SEALER 2-LACQUER			9 LINEAR FT. AT 10⁰⁰/FT.			90⁰⁰		90⁰⁰
									TOTAL	205⁰⁰

Residential Estimate Form

EXTERIOR

Item	Color	Number of coats	Time for setup	Time for prep	Time to paint	Time for cleanup	Total time	Cost of labor	Cost of material	Total price
GRAVEL STOP FASCIA & GUTTERS	D & E TOBACCO BROWN	1-U/C 2-E		1½ HRS.	6 HRS.		7½ HRS.	150⁰⁰	1 GAL-U/C 1 GAL-E = 25⁰⁰	175⁰⁰
4 DOORS	"	1-U/C 2-E		1 HR.	3 HRS.		4 HRS.	80⁰⁰		80⁰⁰
TEXTURED SIDING & GARAGE DOOR	D & E NAVAJO WHITE #60	1-U/C 2-E			3 HRS.		3 HRS.	60⁰⁰	1 GAL U/C 1 GAL-E = 25⁰⁰	85⁰⁰
HAND RAILING	D & E BLACK	1-U/C 2-E		5 MIN	½ HR.		35 MIN	10⁰⁰	1 QT.-E = 5⁰⁰	15⁰⁰
STUCCO	NO PAINT / COLOR COTE									
									TOTAL	355⁰⁰

Completed estimate form
Figure 5-30 (continued)

Paint Shopping List

Brand	Name	Code Number	Flat	Enamel	Quantity
D&E	MELON	67	X		3 GALS.
"	"	"		X	1 GAL.
D&E	SUN TONE	11	X		2 GALS.
"	"	"		X	2 QTS.
D&E	CORK	53	X		2 GALS.
"	"	"		X	2 QTS.
D&E	PEARL WHITE	70	X		6 GALS.
"	"	"		X	4 GALS.
D&E	ALKA-SEALER	28-1			6 GALS.
D&E	U-365 ENAMEL UNDERCOATER	22-1			7 GALS.
D&E	WALNUT STAIN	V-108-22			2 GALS.
D&E	DEC-O-LAC SANDING SEALER	LQ-101			5 GALS.
D&E	DEC-O-LAC SATIN SHEEN	LQ-104			5 GALS.
D&E	TOBACCO BROWN RANCHO HOUSE AND TRIM	60-11		X	1 GAL.
D&E	NAVAJO WHITE RANCHO HOUSE AND TRIM	60-60		X	1 GAL.
D&E	BLACK RANCHO HOUSE AND TRIM	60-20		X	1 QT.

Completed paint shopping list
Figure 5-31

Introduction to Estimating

Equipment Estimate

Item	Description	Cost
SPRAY GUN	CONVENTIONAL - AIR COMPRESSOR	
	RIG WITH LACQUER TIP	100.00 FOR TWO DAYS

Completed equipment estimate
Figure 5-32

Common Materials Estimate

Item	Type	Quantity	Cost
Sandpaper	220	1/2 SLEEVE - 50 SHEETS	$10.00
Sandpaper	80	5 SHEETS	2.00
Spackle	INT.	1 QUART	3.00
Patching material			
Putty			
Caulk	ACRYLIC	3 TUBES	5.00
Paint thinner		5 GALLONS	10.00
Lacquer thinner		5 GALLONS	19.00
Rags	TERRY CLOTH	3 POUNDS	6.00
Roller	3/4" NAP	2	6.00
Roller	1/4" ENAMEL	2	5.00
Tape	1"	2 ROLLS	2.00
Plastic drops	9' X 11'	4	2.00
		TOTAL	$70.00

Completed common materials estimate
Figure 5-33

Estimate Summary

Item	Price
Painting and materials	$2590.00
Common materials	70.00
Equipment	100.00
Subtotal	2760.00
Overhead and profit - _25_ %	690.00
Total	3450.00

Completed estimate summary
Figure 5-34

Introduction to Estimating

Manhours for Painting

The manhour figures in this table should be a useful guide when estimating production rates. These figures assume that labor productivity is fair to good, that working conditions don't delay production unnecessarily, that tools, equipment and materials appropriate for the job are available and used to good advantage, that experienced tradesmen are doing good professional quality painting, and that the job is about the size that most paint contractors are used to handling.

The pages that follow show the production rates I use when estimating. They're valid for most of the work my company handles and for most of the tradesmen on my payroll. Only you can determine whether they're valid for your work and your crews. For example, look at the figure for painting an interior flush door with a roller. Notice that the time listed for putting the first finish coat on one side of the door and one-half of the jamb is .17 hours. That's only 10 minutes (rounded to the nearest whole minute). If your painters take 20 minutes to do that work, either you're doing custom quality painting or your tradesmen are not motivated or skilled enough to handle the production that these manhour tables assume.

Of course, there's nothing wrong with doing custom quality work. My company does all we can handle. But we never do it at production painting prices. Doing custom quality painting at production quality prices is a prescription for disaster. Don't fall into that trap.

Note that all figures are in hours and hundredths of an hour. For example, .25 hours is 15 minutes; .33 hours is 20 minutes. I use decimals of an hour for two reasons. First, adding, subtracting, multiplying and dividing are much easier with hours and decimals of an hour than with hours and minutes. Second, most estimating tables used by professional estimators are in decimals of an hour. Comparing these figures with figures in other sources will be easier because I've expressed times in hours and decimals of an hour.

There aren't many estimating references on painting. One I can recommend is *Estimating Painting Costs* by Dennis Gleason. It's a 450-page manual, fully illustrated, that gives manhour tables for all types of painting, including repaints. It shows how to predict labor productivity, figure labor, material, equipment and subcontractor costs, factor in miscellaneous and contingency costs, adjust markup for the variables in the job, and figure in overhead and profit. Another reference is the *National Construction Estimator*, an annual price book covering the entire construction field, including valuable material for paint contractors. There's an order form at the back of this book for these manuals. A third reference for paint estimators is the 66-page *Estimating Guide* published by the Painting and Decorating Contractors of America, 3913 Old Lee Hwy., Suite 33B, Fairfax, VA 22030.

When using the manhour estimates that follow, please remember that they cover only the actual time spent in painting. Time spent on preparation, set up, clean up, mixing paints, setting up ladders or scaffold or supervising should be added when all production time has been totaled. For most work, non-productive time will be at least 20 percent of the total manhours spent on the job. Don't forget this important part of every estimate.

Manhour tables
Figure 5-35

Interior Painting

This table assumes that standard quality work is being done by experienced painters under normal conditions. The manhour figures listed include no job setup time, no job cleanup time, no surface preparation time, no time spent on hardware removal or replacement and no allowance for supervision. On most jobs setup and cleanup together will add about 10 percent to the time required. Larger jobs and high production work such as apartments will go faster. Smaller jobs and custom work will take longer. Add more time if work is done from ladders. The manhours listed assume a roller is used whenever practical and that little brushwork is needed on walls and ceilings. Spray painting is covered in another table. Paint on windows, doors, trim and cabinets is assumed to be enamel. Paint on walls and ceilings is either enamel or flat.

Sash window with 12 lites. Manhours to paint interior sash and trim with one coat.
Seal or prime unpainted surface	.60 hours
Apply first finish coat	.50 hours
Apply additional finish coat	.40 hours

Sash window with 4 lites or less. Manhours to paint interior sash and trim with one coat.
Seal or prime unpainted surface	.40 hours
Apply first finish coat	.33 hours
Apply additional finish coat	.25 hours

Panel door with a brush. Manhours to paint one side of the door and one-half of the jamb with one coat.
Seal or prime unpainted surface	.60 hours
Apply first finish coat	.50 hours
Apply additional finish coat	.40 hours

Flush door with a brush. Manhours to paint one side of the door and one-half of the jamb with one coat.
Seal or prime unpainted surface	.40 hours
Apply first finish coat	.33 hours
Apply additional finish coat	.25 hours

Flush door with a roller. Manhours to paint one side of the door and one-half of the jamb with one coat. Brush cut-in only.
Seal or prime unpainted surface	.21 hours
Apply first finish coat	.17 hours
Apply additional finish coat	.13 hours

French door with a brush. Manhours to paint one side of a 10 or 15 lite French door and one-half of the jamb with one coat.
Seal or prime unpainted surface	.85 hours
Apply first finish coat	.70 hours
Apply additional finish coat	.55 hours

Louver door with a brush. Manhours to paint one side of the door and one-half of the jamb with one coat.
Seal or prime unpainted surface	.90 hours
Apply first finish coat	.75 hours
Apply additional finish coat	.60 hours

Stair handrail, baluster and newel. Manhours to paint wood rail and spindles with one coat. Based on a flight of stairs with 15 linear feet of handrail and 12 balusters.
Seal or prime unpainted surface	1.20 hours
Apply first finish coat	1.00 hours
Apply additional finish coat	.85 hours

Molding and trim. Linear feet painted per manhour for painting one coat of smooth face base, chair rail, crown, picture molding or window trim. Note that molding painted at the same time as a wall and the same color and finish as the wall takes no more time than painting the wall itself.
Seal or prime unpainted surface	95 LF per hour
Apply first finish coat	120 LF per hour
Apply additional finish coat	145 LF per hour

Manhour tables
Figure 5-35 (continued)

Interior Painting (continued)

Cabinets. Production per manhour for painting one coat on cabinet interior and cabinet face. Use these figures for either 36" high base cabinets or 30" high wall cabinets. No hardware removal or replacement included. Double these figures for full height cabinets from 6' to 8' high. Measure the linear feet of cabinet along the cabinet front or cabinet back, whichever is longer.
 Seal or prime unpainted surface   5.85 LF per hour
 Apply first finish coat   7.50 LF per hour
 Apply additional finish coat   9.15 LF per hour

Bookcases. Production per manhour for painting one coat on an open 6' high bookcase interior and exterior. No disassembly, hardware removal or replacement included. Measure the linear feet of bookcase along the bookcase front or back, whichever is longer.
 Seal or prime unpainted surface   6.25 LF per hour
 Apply first finish coat   8.00 LF per hour
 Apply additional finish coat   9.75 LF per hour

Interior smooth walls. Square feet per manhour for painting one coat with a roller.
 Seal or prime unpainted surface   275 SF per hour
 Apply first finish coat   350 SF per hour
 Apply additional finish coat   425 SF per hour

Interior textured walls. Square feet per manhour for painting one coat with a roller.
 Seal or prime unpainted surface   315 SF per hour
 Apply first finish coat   400 SF per hour
 Apply additional finish coat   490 SF per hour

Interior tongue and groove walls with a brush. Square feet per manhour for painting one coat with a brush when using a roller is not practical.
 Seal or prime unpainted surface   80 SF per hour
 Apply first finish coat   100 SF per hour
 Apply additional finish coat   125 SF per hour

Interior tongue and groove walls with a roller. Square feet per manhour for painting one coat when a roller can be used to good advantage.
 Seal or prime unpainted surface   160 SF per hour
 Apply first finish coat   200 SF per hour
 Apply additional finish coat   250 SF per hour

Smooth ceiling. Square feet per manhour for painting one coat with a roller.
 Seal or prime unpainted surface   215 SF per hour
 Apply first finish coat   275 SF per hour
 Apply additional finish coat   335 SF per hour

Textured ceiling. Square feet per manhour for painting one coat with a roller.
 Seal or prime unpainted surface   245 SF per hour
 Apply first finish coat   315 SF per hour
 Apply additional finish coat   385 SF per hour

Tongue and groove ceiling with a brush. Square feet per manhour for painting one coat with a brush when using a roller is not practical.
 Seal or prime unpainted surface   50 SF per hour
 Apply first finish coat   60 SF per hour
 Apply additional finish coat   75 SF per hour

Tongue and groove ceiling with a roller. Square feet per manhour for painting one coat when a roller can be used to good advantage.
 Seal or prime unpainted surface   100 SF per hour
 Apply first finish coat   120 SF per hour
 Apply additional finish coat   145 SF per hour

Manhour tables
Figure 5-35 (continued)

Exterior Painting

This table assumes that standard quality work is being done by experienced painters under normal conditions. The manhour figures listed include no job setup time, no job cleanup time, no surface preparation time, no time spent moving or protecting trees or shrubs and no allowance for supervision. On most jobs, setup and cleanup together will add about 10 percent to the time required. Larger jobs and high production work such as apartments will go faster. Smaller jobs and custom work will take longer. Except as noted, add more time if work must be done from ladders or scaffolding. The manhours listed for work with a roller includes a minimum of brushwork. Spray painting is covered in another table.

Sash window with 12 lites. Manhours to paint exterior sash and trim with one coat.
Seal or prime unpainted surface	.60 hours
Apply first finish coat	.50 hours
Apply additional finish coat	.40 hours

Sash window with 4 lites or less. Manhours to paint exterior sash and trim with one coat.
Seal or prime unpainted surface	.40 hours
Apply first finish coat	.33 hours
Apply additional finish coat	.25 hours

Panel door with a brush. Manhours to paint one side of the door and one-half of the jamb with one coat.
Seal or prime unpainted surface	.60 hours
Apply first finish coat	.50 hours
Apply additional finish coat	.40 hours

Flush door with a brush. Manhours to paint one side of the door and one-half of the jamb with one coat.
Seal or prime unpainted surface	.40 hours
Apply first finish coat	.33 hours
Apply additional finish coat	.25 hours

Flush door with a roller. Manhours to paint one side of the door and one-half of the jamb with one coat. Brush cut-in only.
Seal or prime unpainted surface	.21 hours
Apply first finish coat	.17 hours
Apply additional finish coat	.13 hours

French door with a brush. Manhours to paint one side of a 10 or 15 lite French door and one-half of the jamb with one coat.
Seal or prime unpainted surface	.80 hours
Apply first finish coat	.70 hours
Apply additional finish coat	.55 hours

Louver door with a brush. Manhours to paint both sides of the door and the complete jamb with one coat.
Seal or prime unpainted surface	.90 hours
Apply first finish coat	.75 hours
Apply additional finish coat	.60 hours

Molding. Linear feet painted per manhour for painting one coat of exterior window or door molding or decorative trim. Note that molding painted at the same time as a wall and the same color and finish as the wall takes no more time than painting the wall itself.
Seal or prime unpainted surface	140 LF per hour
Apply first finish coat	180 LF per hour
Apply additional finish coat	220 LF per hour

Fascia board. Linear feet per manhour for painting one coat with a brush on one face and one edge of 6" to 8" wide fascia. Reduce these figures by one-half if both the interior and exterior face are painted. These figures assume that work is done from a ladder.
Seal or prime unpainted surface	50 LF per hour
Apply first finish coat	60 LF per hour
Apply additional finish coat	70 LF per hour

Manhour tables
Figure 5-35 (continued)

Exterior Painting (continued)

Exterior stairs. Manhours to paint one coat with a brush on treads, risers, handrail, baluster and newel. Based on a flight of 14 treads and 13 risers.
Seal or prime unpainted surface	2.40 hours
Apply first finish coat	2.00 hours
Apply additional finish coat	1.55 hours

Concrete porches and patios. Square feet per manhour for painting one coat with a roller.
Seal or prime unpainted surface	280 SF per hour
Apply first finish coat	350 SF per hour
Apply additional finish coat	420 SF per hour

Wood porches and decks. Square feet per manhour for painting one coat with a roller and brush.
Seal or prime unpainted surface	235 SF per hour
Apply first finish coat	300 SF per hour
Apply additional finish coat	360 SF per hour

Gutters and downspouts. Linear feet per manhour for painting one coat with a brush on the exposed sides of a metal box or half round 5" gutter or 4" downspout. No surface preparation or disassembly included. These figures assume that work is done from a ladder.
Prime unpainted surface	80 LF per hour
Apply first finish coat	100 LF per hour
Apply additional finish coat	120 LF per hour

Exterior shutters. Number of shutters per manhour painting one coat with a brush on both sides and all edges. These figures are based on shutters measuring 2' by 4'. No surface preparation or disassembly included.
Seal or prime unpainted surface	1.75 per hour
Apply first finish coat	2.00 per hour
Apply additional finish coat	2.25 per hour

Wood window screens. Number of screens per manhour painting one coat on all sides. Excludes time required for removing and rehanging screens.
Apply first finish coat	7 per hour
Apply additional finish coat	8 per hour

Wrought iron fence. Linear feet per manhour painting one coat with a brush and a roller. These figures are based on a fence 5' to 6' in height.
Prime unpainted surface	35 LF per hour
Apply first finish coat	40 LF per hour
Apply additional finish coat	50 LF per hour

Exterior smooth walls. Square feet per manhour for painting one coat with a roller.
Seal or prime unpainted surface	275 SF per hour
Apply first finish coat	350 SF per hour
Apply additional finish coat	425 SF per hour

Exterior textured walls. Square feet per manhour for painting one coat with a roller.
Seal or prime unpainted surface	315 SF per hour
Apply first finish coat	400 SF per hour
Apply additional finish coat	490 SF per hour

Exterior tongue and groove walls with a brush. Square feet per manhour for painting one coat with a brush when using a roller is not practical.
Seal or prime unpainted surface	80 SF per hour
Apply first finish coat	100 SF per hour
Apply additional finish coat	125 SF per hour

Exterior tongue and groove walls with a roller. Square feet per manhour for painting one coat when a roller can be used to good advantage.
Seal or prime unpainted surface	160 SF per hour
Apply first finish coat	200 SF per hour
Apply additional finish coat	250 SF per hour

Manhour tables
Figure 5-35 (continued)

Interior Spray Painting

This table assumes that standard quality work is being done by experienced painters under normal conditions. The manhour figures listed include no job setup time, no job cleanup time, no surface preparation time, no time spent on hardware removal or replacement, no time for mixing paint or spray gun maintenance and no allowance for supervision. Larger jobs and high production work such as apartments will go faster. Smaller jobs will take longer. The manhours listed assume little or no hand cut-in. Be aware that on some jobs using a spray gun will take longer than using a brush or roller when the time needed for masking, cleanup of overspray and gun maintenance is considered. These figures assume that the job is well-suited for use of a spray gun.

Panel door. Doors per manhour when spraying one coat on both sides of the door.
 Seal or prime unpainted surface 10 doors
 Apply first finish coat 12 doors
 Apply additional finish coat 14 doors

Flush door. Doors per manhour when spraying one coat on both sides of the door.
 Seal or prime unpainted surface 13 doors
 Apply first finish coat 16 doors
 Apply additional finish coat 19 doors

Louver door. Doors per manhour when spraying one coat on both sides of the door.
 Seal or prime unpainted surface 6 doors
 Apply first finish coat 8 doors
 Apply additional finish coat 10 doors

Stair handrail, baluster and newel. Linear feet of rail per manhour to spray one coat. Based on rail with one baluster for each linear foot of rail.
 Seal or prime unpainted surface 72 LF per hour
 Apply first finish coat 90 LF per hour
 Apply additional finish coat 110 LF per hour

Interior tongue and groove paneling. Square feet per manhour for spraying one coat.
 Seal or prime unpainted surface 465 SF per hour
 Apply first finish coat 600 SF per hour
 Apply additional finish coat 730 SF per hour

Interior smooth walls. Square feet per manhour for spraying one coat.
 Seal or prime unpainted surface 470 SF per hour
 Apply first finish coat 600 SF per hour
 Apply additional finish coat 735 SF per hour

Interior textured walls. Square feet per manhour for spraying one coat.
 Seal or prime unpainted surface 625 SF per hour
 Apply first finish coat 800 SF per hour
 Apply additional finish coat 975 SF per hour

Cabinets. Production per manhour for spraying one coat on cabinet interior and cabinet face. Use these figures for either 36" high base cabinets or 30" high wall cabinets. No hardware removal or replacement included. Double these figures for full height cabinets from 6' to 8' high. Measure the linear feet of cabinet along the cabinet front or cabinet back, whichever is longer.
 Seal or prime unpainted surface 28 LF per hour
 Apply first finish coat 35 LF per hour
 Apply additional finish coat 43 LF per hour

Bookcases. Production per manhour for spraying one coat on an open 6' high bookcase interior and exterior. No disassembly, hardware removal or replacement included. Measure the linear feet of bookcase along the bookcase front or back, whichever is longer.
 Seal or prime unpainted surface 32 LF per hour
 Apply first finish coat 40 LF per hour
 Apply additional finish coat 50 LF per hour

Manhour tables
Figure 5-35 (continued)

Interior Spray Painting (continued)

Smooth ceiling. Square feet per manhour for spraying one coat.
Seal or prime unpainted surface	465 SF per hour
Apply first finish coat	600 SF per hour
Apply additional finish coat	730 SF per hour

Textured ceiling. Square feet per manhour for spraying one coat.
Seal or prime unpainted surface	550 SF per hour
Apply first finish coat	700 SF per hour
Apply additional finish coat	850 SF per hour

Tongue and groove ceiling. Square feet per manhour for spraying one coat.
Seal or prime unpainted surface	390 SF per hour
Apply first finish coat	500 SF per hour
Apply additional finish coat	610 SF per hour

Acoustical "popcorn" ceiling. Square feet per manhour for spraying one coat.
Seal or prime unpainted surface	400 SF per hour
Apply first finish coat	500 SF per hour
Apply additional finish coat	600 SF per hour

Manhour tables
Figure 5-35 (continued)

Paint Contractor's Manual

Exterior Spray Painting

This table assumes that standard quality work is being done by experienced painters under normal conditions. The manhour figures listed include no job setup time, no job cleanup time, no surface preparation time, no time spent on hardware removal or replacement, no time for mixing paint or spray gun maintenance and no allowance for supervision. Larger jobs and high production work such as apartments will go faster. Smaller jobs will take longer. The manhours listed assume little or no hand cut-in. Be aware that on some jobs using a spray gun will take longer than using a brush or roller when the time needed for masking, cleanup of overspray and gun maintenance is considered. These figures assume that the job is well-suited for use of a spray gun.

Panel door. Doors per manhour when spraying one coat on one side of the door.
Seal or prime unpainted surface	18 doors
Apply first finish coat	20 doors
Apply additional finish coat	23 doors

Flush door. Doors per manhour when spraying one coat on one side of the door.
Seal or prime unpainted surface	20 doors
Apply first finish coat	26 doors
Apply additional finish coat	32 doors

Louver door. Doors per manhour when spraying one coat on one side of the door.
Seal or prime unpainted surface	11 doors
Apply first finish coat	14 doors
Apply additional finish coat	17 doors

Fascia board. Linear feet per manhour for spraying one coat on one face of 6" to 8" wide fascia. These figures assume that work is done from a ladder.
Seal or prime unpainted surface	280 LF per hour
Apply first finish coat	360 LF per hour
Apply additional finish coat	440 LF per hour

Concrete porches and patios. Square feet per manhour for spraying one coat.
Seal or prime unpainted surface	1400 SF per hour
Apply first finish coat	1800 SF per hour
Apply additional finish coat	2200 SF per hour

Open soffit. Square feet per manhour for spraying one coat on the underside of roof sheathing, both sides of rafters and the interior of the fascia. Low and wide soffits will take less time. Narrow and high soffits will take longer. These figures assume that work is done from a ladder.
Seal or prime unpainted surface	280 SF per hour
Apply first finish coat	360 SF per hour
Apply additional finish coat	450 SF per hour

Exterior stairs. Manhours to spray one coat on treads, risers, handrail and newel posts. Based on a flight of 14 treads and 13 risers.
Seal or prime unpainted surface	.60 hours
Apply first finish coat	.50 hours
Apply additional finish coat	.40 hours

Exterior shutters. Number of shutters per manhour spraying one coat on both sides. These figures are based on shutters measuring 2' by 4'. No surface preparation or disassembly included.
Seal or prime unpainted surface	9.25 per hour
Apply first finish coat	12.00 per hour
Apply additional finish coat	14.50 per hour

Wrought iron fence. Linear feet per manhour spraying one coat. These figures are based on fences 5' to 6' in height.
Prime unpainted surface	190 LF per hour
Apply first finish coat	240 LF per hour
Apply additional finish coat	290 LF per hour

Manhour tables
Figure 5-35 (continued)

Exterior Spray Painting (continued)

Wood porches and decks. Square feet per manhour for spraying one coat.
Seal or prime unpainted surface 1170 SF per hour
Apply first finish coat 1500 SF per hour
Apply additional finish coat 1850 SF per hour

Shingle siding. Square feet per manhour for spraying one coat.
Seal or prime unpainted surface 1100 SF per hour
Apply first finish coat 1400 SF per hour
Apply additional finish coat 1700 SF per hour

Exterior stucco. Square feet per manhour for spraying one coat.
Seal or prime unpainted surface 1400 SF per hour
Apply first finish coat 1800 SF per hour
Apply additional finish coat 2200 SF per hour

Smooth brick or block. Square feet per manhour for spraying one coat.
Seal or prime unpainted surface 1250 SF per hour
Apply first finish coat 1600 SF per hour
Apply additional finish coat 1950 SF per hour

Smooth face siding. Square feet per manhour for spraying one coat.
Seal or prime unpainted surface 935 SF per hour
Apply first finish coat 1200 SF per hour
Apply additional finish coat 1450 SF per hour

Rough brick or block. Square feet per manhour for spraying one coat.
Seal or prime unpainted surface 800 SF per hour
Apply first finish coat 1000 SF per hour
Apply additional finish coat 1220 SF per hour

Rough sawn siding. Square feet per manhour for spraying one coat.
Seal or prime unpainted surface 1100 SF per hour
Apply first finish coat 1400 SF per hour
Apply additional finish coat 1700 SF per hour

Smooth concrete wall. Square feet per manhour for spraying one coat.
Seal or prime unpainted surface 1250 SF per hour
Apply first finish coat 1600 SF per hour
Apply additional finish coat 1950 SF per hour

Stripping Paint

The time required to remove paint, varnish, urethane (or polyurethane), lacquer or stain will vary with the type of coating used and the surface that's been coated. There's no reliable way to estimate this work. Your best bet is to charge by the manhour and for the materials used.

Doors, cabinet faces, drawers and other small items can be sent to a professional strip shop that handles this type of work at a set price. If you use a strip shop, charge the customer enough to cover the cost of removal, transportation to the stripper and back and reinstallation of each item.

Any time liquid stripper is used on wood veneer, there's a good chance that the veneer will lift or be damaged in the process. Stripping solution tends to dissolve the glue used to bond veneer. Be sure your client understands that some damage is likely, even when stripping is done by experienced professional craftsmen. De-laminated veneer can be repaired, of course, but this is a cost your customer should be ready to bear. Your contract should make it clear that stripping is done at the client's risk.

Manhour tables
Figure 5-35 (continued)

Paint Contractor's Manual

Interior and Exterior Staining

This table assumes that standard quality staining is being done by experienced tradesmen under normal conditions. The manhour figures listed include no job setup time, no job cleanup time, no surface preparation time, no time spent on hardware removal or replacement and no allowance for supervision. On most jobs, setup and cleanup together will add about 10 percent to the time required. Fine custom stainwork will take longer. Except as noted, application is with a brush and rag and wipeoff is with a rag.

Sash window with 12 lites. Manhours to stain one side of a window with one coat. Includes sash and trim.
  Seal unpainted surface   .20 hours
  Apply stain coat   .33 hours

Sash window with 4 lites or less. Manhours to stain one side of a window with one coat. Includes sash and trim.
  Seal unpainted surface   .15 hours
  Apply stain coat   .25 hours

Panel door. Manhours to stain one side of the door and one-half of the jamb with one coat.
  Seal unpainted surface   .20 hours
  Apply stain coat   .33 hours

Flush door. Manhours to stain one side of the door and one-half of the jamb with one coat.
  Seal unpainted surface   .15 hours
  Apply stain coat   .25 hours

French door. Manhours to stain one side of a 10 or 15 lite French door and one-half of the jamb with one coat.
  Seal unpainted surface   .30 hours
  Apply stain coat   .50 hours

Louver door. Manhours to stain both sides of the door and the complete jamb with one coat.
  Seal unpainted surface   .30 hours
  Apply stain coat   .50 hours

Stair handrail, baluster and newel. Linear feet of rail per manhour. Includes staining both rail and spindles spaced at 12" along the rail.
  Seal unpainted surface   48 LF per hour
  Apply stain coat   30 LF per hour

Molding and trim. Linear feet per manhour for staining one coat of smooth face base, chair rail, crown, picture molding or window trim.
  Seal unpainted surface   310 LF per hour
  Apply stain coat   200 LF per hour

Cabinets. Production per manhour for staining one coat on cabinet interior and cabinet face. Use these figures for either 36" high base cabinets or 30" high wall cabinets. No hardware removal or replacement included. Double these figures for full height cabinets from 6' to 8' high. Measure the linear feet of cabinet along the cabinet front or cabinet back, whichever is longer.
  Seal unpainted surface   24 LF per hour
  Apply stain coat   15 LF per hour

Manhour tables
Figure 5-35 (continued)

Introduction to Estimating

Interior and Exterior Staining (continued)

Bookcases. Production per manhour for staining one coat on an open 6' high bookcase interior and exterior. No disassembly, hardware removal or replacement included. Measure the linear feet of bookcase along the bookcase front or back, whichever is longer.

Seal unpainted surface	26 LF per hour
Apply stain coat	16 LF per hour

Interior paneling. Square feet per manhour for staining one coat.

Seal unpainted surface	240 SF per hour
Apply stain coat	150 SF per hour

Tongue and groove ceiling with wood beams. Square feet per manhour for spraying one coat.

Seal unpainted surface	310 SF per hour
Apply stain coat	200 SF per hour

Smooth face siding. Square feet per manhour for spraying one coat.

Seal unpainted surface	940 SF per hour
Apply stain coat	600 SF per hour

Rough sawn siding. Square feet per manhour for spraying one coat.

Seal unpainted surface	1250 SF per hour
Apply stain coat	800 SF per hour

Shingle siding. Square feet per manhour for spraying one coat.

Seal unpainted surface	1250 SF per hour
Apply stain coat	800 SF per hour

Wood deck with railing. Square feet per manhour for spraying one coat.

Seal unpainted surface	2340 SF per hour
Apply stain coat	1500 SF per hour

Varnishing

Applying varnish or urethane takes about the same time as doing quality work with enamel. Use the interior painting table or interior spray painting table when estimating varnish application. But there's an extra step in varnishing. The surface should be rubbed with steel wool to remove any roughness. Figure 200 square feet per manhour for buffing a varnished surface with steel wool. If the surface has to be both buffed with steel wool and waxed, figure production at 100 square feet per manhour.

Manhour tables
Figure 5-35 (continued)

Lacquering

This table assumes that standard quality lacquering is being done by experienced tradesmen under normal conditions. The manhour figures listed include no job setup time, no job cleanup time, no surface preparation time, no time spent on hardware removal or replacement and no allowance for supervision. On most jobs setup and cleanup together will add about 10 percent to the time required. Fine custom work will take longer than indicated here. No time is included in this table for the initial staining. Add staining time from the table for interior and exterior staining.

Panel door. Estimate .50 manhours for one side of the door. This includes two coats of sanding sealer, sanding by hand, putty, dusting and two coats of finish lacquer.

Flush door. Estimate .42 manhours for one side of the door. This includes two coats of sanding sealer, sanding by hand, putty, dusting and two coats of finish lacquer.

Louver door. Estimate 1.5 manhours for both sides of the door. This includes two coats of sanding sealer, sanding by hand, putty, dusting and two coats of finish lacquer.

Stair rail. Estimate 2 manhours for 15 feet of handrail and 15 balusters. This includes two coats of sanding sealer, sanding by hand, putty, dusting and two coats of finish lacquer.

Molding and trim. Estimate 50 linear feet per manhour. This includes two coats of sanding sealer, sanding by hand, putty, dusting and two coats of finish lacquer. Use these figures for smooth face base, chair rail, crown, picture molding and window trim.

Paneling. Estimate 75 square feet per manhour. This includes two coats of sanding sealer, sanding by hand, putty, dust and two coats of finish lacquer.

Cabinets. Estimate 5 linear feet per manhour when lacquering doors, stiles and interior shelves. This includes two coats of sanding sealer, sanding by hand, putty, dusting, and two coats of finish lacquer. Use these figures for either 36" high base cabinets or 30" high wall cabinets. No hardware removal or replacement is included. Full height cabinets from 6' to 8' high will cut the production rate by one-half. Measure the linear feet of cabinet along the cabinet front or cabinet back, whichever is longer.

Cabinet faces and stiles only. Estimate 7 linear feet per manhour when lacquering doors and stiles only. This includes two coats of sanding sealer, sanding by hand, putty, dusting, and two coats of finish lacquer. Use these figures for either 36" high base cabinets or 30" high wall cabinets. No hardware removal or replacement is included. Full height cabinets from 6' to 8' high will cut the production rate by one-half. Measure the linear feet of cabinet along the cabinet front or cabinet back, whichever is longer.

Bookcases. Estimate 5 linear feet per manhour when lacquering the interior and exterior of 6' high bookcases if no disassembly, hardware removal or replacement is needed. This includes two coats of sanding sealer, sanding by hand, putty, dusting, and two coats of finish lacquer. Measure the linear feet of bookcase along the case front or back, whichever is longer.

Manhour tables
Figure 5-35 (continued)

Introduction to Estimating

Interior Ceilings and Walls Checklist

Painting

	Room and number	Area	No. coats	Color	Ceiling type
☐	1 _____	_____	_____	_____	_____
☐	2 _____	_____	_____	_____	_____
☐	3 _____	_____	_____	_____	_____
☐	4 _____	_____	_____	_____	_____
☐	5 _____	_____	_____	_____	_____
☐	6 _____	_____	_____	_____	_____
☐	7 _____	_____	_____	_____	_____
☐	8 _____	_____	_____	_____	_____
☐	9 _____	_____	_____	_____	_____
☐	10 _____	_____	_____	_____	_____

Surface Preparation

	Item	Room numbers	Quantity	Comment
☐	Cleaning	_____	_____	_____
☐	Repair cracks	_____	_____	_____
☐	Repair holes	_____	_____	_____
☐	Scraping	_____	_____	_____
☐	Sanding	_____	_____	_____
☐	Repair flaking	_____	_____	_____
☐	Caulking	_____	_____	_____
☐	Sealing	_____	_____	_____
☐	Repair tape marks	_____	_____	_____
☐	Repair water damage	_____	_____	_____
☐	_____	_____	_____	_____
☐	_____	_____	_____	_____

Materials

Paint	Color	Area	Coverage	Quantity
_____	_____	_____	_____	_____
_____	_____	_____	_____	_____
_____	_____	_____	_____	_____
Primer		_____	_____	_____
Primer		_____	_____	_____
Masking tape		_____	_____	_____
Miscellaneous prep materials		_____	_____	_____
Other materials		_____	_____	_____

Setup and Cleanup

Item	Calculation or comment
Move and protect furniture	_____
Cover floor	_____
Protect carpet	_____
Cover windows and mirrors	_____
Set up each day	_____
Clean up each day	_____
_____	_____
_____	_____

Interior Doors and Jambs Checklist

Painting

Type	Number of doors	Number of coats	Color	Method of application
☐ Flush	_____	_____	_____	_____
☐ Inset panel	_____	_____	_____	_____
☐ Louvered	_____	_____	_____	_____
☐ Pocket sliding	_____	_____	_____	_____
☐ Metal	_____	_____	_____	_____
☐ Closet	_____	_____	_____	_____
☐ _____	_____	_____	_____	_____
☐ _____	_____	_____	_____	_____

Surface Preparation

Item	Quantity	Comment
☐ Removing hardware	_____	_____
☐ Removing doors	_____	_____
☐ Cleaning	_____	_____
☐ Scraping	_____	_____
☐ Sanding	_____	_____
☐ Patching	_____	_____
☐ Caulking	_____	_____
☐ Priming	_____	_____
☐ Rehanging doors	_____	_____
☐ Replacing hardware	_____	_____
☐ _____	_____	_____
☐ _____	_____	_____

Materials

Paint	Color	Area	Coverage	Quantity
_____	_____	_____	_____	_____
_____	_____	_____	_____	_____
_____	_____	_____	_____	_____
_____	_____	_____	_____	_____
Primer		_____	_____	_____
Primer		_____	_____	_____
Masking tape		_____	_____	_____
Miscellaneous prep materials		_____	_____	_____
Other materials		_____	_____	_____

Setup and Cleanup

Item	Calculation or comment
Protect wallpaper, etc.	_____
Spray gun maintenance	_____
Mask off carpet	_____
Set up each day	_____
Clean up each day	_____
_____	_____
_____	_____
_____	_____

Introduction to Estimating

Interior Windows and French Doors Checklist

Painting

Type	Number of windows	Size	Number of lites	Number of coats	Color
☐ Aluminum	_____	_____	_____	_____	_____
☐ Wood	_____	_____	_____	_____	_____
☐ Bay	_____	_____	_____	_____	_____
☐ Casement	_____	_____	_____	_____	_____
☐ Louvered	_____	_____	_____	_____	_____
☐ French	_____	_____	_____	_____	_____
☐ Fixed	_____	_____	_____	_____	_____
☐ Shutters	_____	_____	_____	_____	_____
☐ _____	_____	_____	_____	_____	_____

Surface Preparation

Item	Number of windows	Comment
☐ Removing hardware	_____	_____
☐ Cleaning	_____	_____
☐ Scraping	_____	_____
☐ Sanding	_____	_____
☐ Patching	_____	_____
☐ Caulking	_____	_____
☐ Priming	_____	_____
☐ Rehanging windows	_____	_____
☐ Replacing hardware	_____	_____
☐ Glazing	_____	_____
☐ _____	_____	_____

Materials

Paint	Color	Linear Feet	Coverage	Quantity
_____	_____	_____	_____	_____
_____	_____	_____	_____	_____
_____	_____	_____	_____	_____
_____	_____	_____	_____	_____
_____	_____	_____	_____	_____
Primer		_____	_____	_____
Primer		_____	_____	_____
Masking tape		_____	_____	_____
Miscellaneous prep materials		_____	_____	_____
Other materials		_____	_____	_____

Setup and Cleanup

Item	Calculation or comment
Protect adjacent surfaces	_____
Masking	_____
Set up each day	_____
Clean up each day	_____
_____	_____
_____	_____

Paint Contractor's Manual

Interior Trim Checklist

Painting

Type	Length or area	Number of coats	Color	Method
☐ Baseboard	_____	_____	_____	_____
☐ Crown	_____	_____	_____	_____
☐ Ceiling beams	_____	_____	_____	_____
☐ Chair rail	_____	_____	_____	_____
☐ Shutters	_____	_____	_____	_____
☐ Wainscot	_____	_____	_____	_____
☐ _____	_____	_____	_____	_____
☐ _____	_____	_____	_____	_____

Surface Preparation

Item	Quantity	Comment
☐ Removing hardware	_____	_____
☐ Removing trim	_____	_____
☐ Cleaning	_____	_____
☐ Scraping	_____	_____
☐ Sanding	_____	_____
☐ Patching	_____	_____
☐ Caulking	_____	_____
☐ Priming	_____	_____
☐ Replacing trim	_____	_____
☐ Replacing hardware	_____	_____
☐ _____	_____	_____
☐ _____	_____	_____

Materials

Paint	Color	Area	Coverage	Quantity
_____	_____	_____	_____	_____
_____	_____	_____	_____	_____
_____	_____	_____	_____	_____
_____	_____	_____	_____	_____
Scaffolding		_____	_____	_____
Primer		_____	_____	_____
Primer		_____	_____	_____
Masking tape		_____	_____	_____
Miscellaneous prep materials		_____	_____	_____
Other materials		_____	_____	_____

Setup and Cleanup

Item	Calculation or comment
Protect wallpaper	_____
Spray gun maintenance	_____
Mask off carpet	_____
Set up each day	_____
Clean up each day	_____
_____	_____
_____	_____
_____	_____
_____	_____

Introduction to Estimating

Interior Cabinets Checklist

Painting

Type	Number or area	Number of coats	Color	Method
☐ Door type:_____	_____	_____	_____	_____
☐ Door type:_____	_____	_____	_____	_____
☐ Door type:_____	_____	_____	_____	_____
☐ Shelves	_____	_____	_____	_____
☐ Interiors	_____	_____	_____	_____
☐ _____	_____	_____	_____	_____
☐ _____	_____	_____	_____	_____
☐ _____	_____	_____	_____	_____

Surface Preparation

Item	Quantity	Comment
☐ Removing hardware	_____	_____
☐ Removing doors	_____	_____
☐ Cleaning	_____	_____
☐ Scraping	_____	_____
☐ Sanding	_____	_____
☐ Patching	_____	_____
☐ Caulking	_____	_____
☐ Priming	_____	_____
☐ Rehanging doors	_____	_____
☐ Replacing hardware	_____	_____
☐ _____	_____	_____
☐ _____	_____	_____

Materials

Paint	Color	Area	Coverage	Quantity
_____	_____	_____	_____	_____
_____	_____	_____	_____	_____
_____	_____	_____	_____	_____
Primer		_____	_____	_____
Primer		_____	_____	_____
Masking tape		_____	_____	_____
Miscellaneous prep materials		_____	_____	_____
Other materials		_____	_____	_____

Setup and Cleanup

Item	Calculation or comment
Remove contents	_____
Cover floor	_____
Cover counter tops	_____
Set up each day	_____
Clean up each day	_____
_____	_____
_____	_____
_____	_____
_____	_____
_____	_____

Interior Stripping Checklist

Painting

Description	Area	Paint to remove	Paint thickness	Stain or repaint with:
☐				
☐				
☐				
☐				
☐				
☐				
☐				
☐				

Surface Preparation

	Item	Quantity	Comment
☐	Removing hardware		
☐	Removing doors		
☐	Scraping		
☐	Sanding		
☐	Masking		
☐	Bleaching		
☐	Rehanging doors		
☐	Replacing hardware		
☐			
☐			

Materials

Material	Description	Area	Coverage	Quantity
Stripper				
Stripper				
Acetone				
M.E.K.				
T.S.P.				
Bleach				
Masking tape				
Miscellaneous prep materials				
Other materials				

Setup and Cleanup

Item	Calculation or comment
Protect wallpaper	
Water cleanup	
Mask off carpet	
Set up each day	
Clean up each day	

Introduction to Estimating

Interior Staining and Finishing Checklist

Staining

Type	Area	Number of coats	Color	Method of application
☐ Softwood	_____	_____	_____	_____
☐ Hardwood	_____	_____	_____	_____
☐ New wood	_____	_____	_____	_____
☐ Stripped wood	_____	_____	_____	_____
☐ _____	_____	_____	_____	_____
☐ _____	_____	_____	_____	_____

Surface Preparation

Item	Quantity	Comment
☐ Removing hardware	_____	_____
☐ Removing doors	_____	_____
☐ Sanding	_____	_____
☐ Filling nail holes	_____	_____
☐ Sealing	_____	_____
☐ Rehanging doors	_____	_____
☐ Replacing hardware	_____	_____
☐ Wet and dry sanding	_____	_____
☐ _____	_____	_____

Materials

Coating	Color	Area	Coverage	Quantity
Stain	_____	_____	_____	_____
Varnish	_____	_____	_____	_____
Lacquer	_____	_____	_____	_____
Urethane	_____	_____	_____	_____
Steel wool	_____	_____	_____	_____
Buffing compound	_____	_____	_____	_____
Wax	_____	_____	_____	_____
Masking tape	_____	_____	_____	_____
Miscellaneous prep materials	_____	_____	_____	_____
Other materials	_____	_____	_____	_____

Setup and Cleanup

Item	Calculation or comment
Protect wallpaper	_____
Spray gun maintenance	_____
Mask off carpet	_____
Set up each day	_____
Clean up each day	_____
_____	_____
_____	_____
_____	_____
_____	_____
_____	_____

Paint Contractor's Manual

Exterior Siding Checklist

Painting

	Wall or item	Area	No. coats	Color	Method
☐	1_____	_____	_____	_____	_____
☐	2_____	_____	_____	_____	_____
☐	3_____	_____	_____	_____	_____
☐	4_____	_____	_____	_____	_____
☐	5_____	_____	_____	_____	_____
☐	6_____	_____	_____	_____	_____
☐	7_____	_____	_____	_____	_____
☐	8_____	_____	_____	_____	_____
☐	9_____	_____	_____	_____	_____
☐	10_____	_____	_____	_____	_____

Surface Preparation

	Item	Wall numbers	Quantity	Comment
☐	Cleaning	_____	_____	_____
☐	Water blast	_____	_____	_____
☐	Scraping	_____	_____	_____
☐	Fill cracks	_____	_____	_____
☐	Repair holes	_____	_____	_____
☐	Sanding	_____	_____	_____
☐	Caulking	_____	_____	_____
☐	Priming	_____	_____	_____
☐	Repair water damage	_____	_____	_____
☐	_____	_____	_____	_____
☐	_____	_____	_____	_____

Materials

Paint	Color	Area	Coverage	Quantity
_____	_____	_____	_____	_____
_____	_____	_____	_____	_____
_____	_____	_____	_____	_____
Primer			_____	_____
Primer			_____	_____
Masking tape			_____	_____
Patching			_____	_____
Water blaster			_____	_____
Disc sander			_____	
Scaffolding				

Setup and Cleanup

Item	Calculation or comment
Trim trees, shrubbery, etc.	_____
Protect landscaping	_____
Erect scaffold	_____
Dismantle scaffold	_____
Cover windows or trim	_____
Spray gun maintenance	_____
Remove overspray	_____
Set up each day	_____
Clean up each day	_____
_____	_____
_____	_____

Introduction to Estimating

Exterior Stucco Checklist

Painting

Wall or item number	Area	No. coats	Color	Method
☐ 1				
☐ 2				
☐ 3				
☐ 4				
☐ 5				
☐ 6				
☐ 7				
☐ 8				
☐ 9				
☐ 10				

Surface Preparation

Item	Wall numbers	Quantity	Comment
☐ Found B.C.P.			
☐ Found excess chalkiness			
☐ Repair cracks			
☐ Repair holes			
☐ Scraping			
☐ Sandblast			
☐ Cleaning			
☐ Repair flaking			
☐ Caulking			
☐ Sealing			
☐ Correct moisture problem			
☐ Repair water damage			
☐ Water blast			
☐			

Materials

Paint	Color	Area	Coverage	Quantity
Primer				
Primer				
Masking tape				
Patching materials				
Scaffold				
Water blaster				

Setup and Cleanup

Item	Calculation or comment
Trim trees, shrubbery, etc.	
Protect landscaping	
Erect scaffold	
Dismantle scaffold	
Cover windows or trim	
Set up each day	
Clean up each day	
Spray gun maintenance	
Remove overspray	

Paint Contractor's Manual

Exterior Doors and Jambs Checklist

Painting

Type	Number of doors	Number of coats	Color	Method of application
☐ Flush	_____	_____	_____	_____
☐ Inset panel	_____	_____	_____	_____
☐ Inset glass	_____	_____	_____	_____
☐ Sliding	_____	_____	_____	_____
☐ Screen	_____	_____	_____	_____
☐ Garage	_____	_____	_____	_____
☐ _____	_____	_____	_____	_____
☐ _____	_____	_____	_____	_____

Surface Preparation

Item	Quantity	Comment
☐ Removing hardware	_____	_____
☐ Removing doors	_____	_____
☐ Cleaning	_____	_____
☐ Scraping	_____	_____
☐ Sanding	_____	_____
☐ Patching	_____	_____
☐ Caulking	_____	_____
☐ Priming	_____	_____
☐ Rehanging doors	_____	_____
☐ Replacing hardware	_____	_____
☐ _____	_____	_____
☐ _____	_____	_____

Materials

Paint	Color	Area	Coverage	Quantity
_____	_____	_____	_____	_____
_____	_____	_____	_____	_____
_____	_____	_____	_____	_____
Primer				_____
Primer				_____
Masking tape				_____
Miscellaneous prep materials				_____
Other materials				_____

Setup and Cleanup

Item	Calculation or comment
Trim trees, shrubbery, etc.	_____
Protect landscaping	_____
Spray gun maintenance	_____
Scrape glass	_____
Remove overspray	_____
Set up each day	_____
Clean up each day	_____
_____	_____
_____	_____
_____	_____

Introduction to Estimating

Exterior Windows and French Doors Checklist

Painting

Type	Number of windows	Size	Number of lites	Number of coats	Color
☐ Aluminum					
☐ Wood					
☐ Awning					
☐ Bay					
☐ Casement					
☐ Louvered					
☐ French					
☐ Fixed					
☐ Shutters					
☐ _____					

Surface Preparation

Item	Number of windows	Comment
☐ Unsticking		
☐ Removing hardware		
☐ Removing shutters		
☐ Cleaning		
☐ Scraping		
☐ Sanding		
☐ Patching		
☐ Caulking		
☐ Priming		
☐ Rehanging shutters		
☐ Replacing hardware		
☐ Glazing		
☐ _____		

Materials

Paint	Color	Area	Coverage	Quantity
Primer				
Primer				
Masking tape				
Miscellaneous prep materials				
Other materials				

Setup and Cleanup

Item	Calculation or comment
Trim trees, shrubbery, etc.	
Protect landscaping	
Scrape glass	
Masking	
Set up each day	
Clean up each day	
_____	
_____	

Exterior Trim Checklist

Painting

Type	Length or area	Number of coats	Color	Method
☐ Gutters	_____	_____	_____	_____
☐ Downspouts	_____	_____	_____	_____
☐ Metal trim	_____	_____	_____	_____
☐ Wood trim	_____	_____	_____	_____
☐ Sheet metal	_____	_____	_____	_____
☐ Fencing	_____	_____	_____	_____
☐ Patio floor	_____	_____	_____	_____
☐ _____	_____	_____	_____	_____
☐ _____	_____	_____	_____	_____

Surface Preparation

Item	Quantity	Comment
☐ Removing hardware	_____	_____
☐ Removing trim	_____	_____
☐ Cleaning	_____	_____
☐ Scraping rust	_____	_____
☐ Sanding	_____	_____
☐ Patching	_____	_____
☐ Caulking	_____	_____
☐ Priming	_____	_____
☐ Replacing trim	_____	_____
☐ Replacing hardware	_____	_____
☐ _____	_____	_____
☐ _____	_____	_____

Materials

Paint	Color	Area	Coverage	Quantity
_____	_____	_____	_____	_____
_____	_____	_____	_____	_____
_____	_____	_____	_____	_____
_____	_____	_____	_____	_____
Scaffolding		_____	_____	_____
Primer		_____	_____	_____
Primer		_____	_____	_____
Masking tape		_____	_____	_____
Miscellaneous prep materials		_____	_____	_____
Other materials		_____	_____	_____

Setup and Cleanup

Item	Calculation or comment
Trim trees, shrubbery, etc.	_____
Protect landscaping	_____
Spray gun maintenance	_____
Set up each day	_____
Clean up each day	_____
Remove overspray	_____
_____	_____
_____	_____

Chapter 6

Planning the Job

If your company is organized and staffed, if the public knows you're in business, if you're getting inquiries and preparing accurate estimates, if your estimates are bringing in good jobs that can earn good profits, you're well on the way to building a successful painting business.

But lining up good, profitable work isn't the same as making good profits. It takes both careful planning and good follow-through before any of those profits hit your bank account. Planning the job is the subject of this chapter. Well-planned jobs are quicker, smoother and more profitable. The next two chapters will explain how to do high-production painting.

This chapter will help you plan and organize the job. Even small jobs go faster when they're planned and organized before work begins. And planning is absolutely essential on bigger jobs. No matter how small your company is now, make good planning a habit. It's a habit that makes future expansion possible.

Work Schedule

Scheduling is going to be a hit-and-miss proposition if your company doesn't have a work schedule. It's important for both you *and* your crew. The schedule should establish when you start your day and when you end it. Without a clear work schedule, the boss will end up working nights and weekends while crew members come and go as they please.

Of course, there will be times when you have to deviate from the schedule. Little emergencies are common in every business. But the more you stick to a schedule, the more accepted that schedule becomes and the more productive your crews will be during work hours.

Your crew should have a specific time to begin work — including a starting time on weekends if weekend work is needed. The starting time isn't the time when everyone shows up, starts drinking coffee and discussing last night's T.V. programs. The starting time is when paint cans are opened and brushes start laying down a coating. If you have employees that have to talk for 30 minutes before they start work, ask them to show up for work 30 minutes early.

It's up to you, the boss, to set the work schedule. Make sure everyone knows that schedule and make sure it's enforced. Employees are expected to work these hours *every day* unless they notify you in advance that they need time off. Make it clear how many breaks are allowed and how long breaks should last.

Sample Work Schedule for All Crews

8:00 A.M.	Begin work on site
12:00 to 12:30 P.M.	Lunch
12:30 to 3:15 P.M.	Work continues
3:15 to 3:30 P.M.	Clean up
3:30 P.M.	Quitting time

A simple schedule like this leaves no doubt about what's expected. It also makes possible accurate predictions of the work to be done each day. A work schedule is the most basic planning tool. *You just can't afford to let your employees set their own schedules.*

You, the contractor, will follow a different schedule. Your schedule will probably change daily as requirements change. I usually spend the last 15 minutes or so of each working day thinking through the next day's work. I make up in my mind a simple schedule of the things that have to be done the next day and when I plan to do them:

7:00 A.M. — Pick up supplies for Smith and King jobs at the paint store.

8:00 A.M. — Arrive at Smith job. Get the crew working. Discuss work with Mr. Smith and pick up check for interior work.

8:30 A.M. — Check on progress at King job. Enough paint on hand to finish the exterior?

9:00 A.M. — Estimate laundry redo at 1200 block on South 14th Street.

10:00 A.M. — Return to office, make phone calls, handle mail and paperwork.

12:00 noon — Lunch

12:30 P.M. — Prepare estimate for the laundry job and call in bid total.

1:00 P.M. — Return to King job and paint until 3:30.

4:30 P.M. — Pick up plans for The Meadows tract and prepare estimate.

Any contractor who's meeting the public and scheduling appointments needs a date book to note important due dates and meetings. If you don't already have one, buy an appointment book that covers a full year. Use it to note meetings, tax payment dates, the start and completion date on every job, and anything important enough to warrant preserving a record.

The Role of a Foreman

Every crew needs a team leader — someone who is there all the time that work is going on. I'll call this team leader the foreman. He's a painter, probably the most experienced and most productive on the job. But he's also a supervisor. He takes primary responsibility for supervising production. That requires good planning, logical thinking, good judgement, and initiative.

In a smaller painting company, the contractor is his own foreman. In a larger company, a foreman will be assigned to each crew.

The foreman's primary business is keeping the crew busy and working. On a larger project, there's a lot happening at once. One room is being prepared for painting. Another is in the final painting stages. Meanwhile the last coat of sealer is drying on the handrails. Painters working independently without coordination sometimes work at cross-purposes. A foreman keeps this from happening.

It takes a lot of planning by a foreman to keep the crew producing all day. If no one plans the work, you'll arrive at the job to find the entire crew sitting in the shade waiting for some paint to dry. Your only choice then is either to pay your men for waiting or release the crew for the day. Neither is good for business.

Before work starts, the foreman plans how it will be done and in what order. The order in which rooms are painted may be important to your client. Most homeowners want bedrooms painted first so they'll be useable as soon as possible. If the client has specific ideas on the order of work, let the foreman know about the request. The person who sold the job or made the estimate knows what the client wants and should tell the foreman.

The foreman also has to find out what other tradesmen will be on the job. Coordination with other trades is essential. But dealing with other tradesmen and their schedules can be the biggest scheduling problem. If other construction work is going on during the painting, it's common for many trades to be scheduled simultaneously.

Painting is usually done last. At this stage of construction there's generally an irresistible urge to hurry and complete the job as soon as possible. Sometimes it's because the project is already running late. Sometimes it's because the builder is running out of cash and needs the final payment. In any event, many tradesmen may be working while painting is going on.

If you run into one of these jobs, the only way to handle it is to talk to the other tradesmen. Find out when the carpenter will be finished hanging doors. Ask the tile man when he plans to set the tile in the bathroom. Are the electricians finished? (You don't need them running wire through a wall you just finished painting.)

If the foreman doesn't stay on top of all of this, he'll waste time repainting doors or have a crew waiting most of a day for the plumbers to finish.

The key here is planning and communicating. The foreman has to plan a logical sequence of work and then be sure that sequence doesn't conflict with some other trade's schedule.

Job Work Order

Introduce your foreman to every job with the Job Work Order. See Figure 6-1. It identifies the job and lists several important details the foreman needs to know: the person to contact at the job site, the estimated start and completion dates, the total estimated manhours. You may want to include a short summary of the job and probably will have some special instructions the foreman has to follow.

Attach the Job Work Order to a copy of the original cost estimate and a copy of the cover letter that was sent to the client with the estimate. Give this packet of documents to the foreman several days before his crew is assigned to the job. These documents provide all the information the foreman needs to complete the work. The estimate defines the scope of the work and identifies exactly what has to be painted. The cover letter describes the paint and color to be applied to each surface. The Job Work Order includes room to list special instructions or other directions that may be needed.

If your foreman is going to pick up supplies for this job, he also needs the Paint Shopping List. On new construction, your foreman will probably want to have the set of plans you bid from, if those plans are available.

The Foreman's Plan

Once the foreman knows the details of the job and the date work is to begin, he can start planning. He begins with a concept plan for the entire job. Then he develops day-by-day plans that cover the work to be done each day. Before the job begins, he should have a plan like this:

Work Summary

Seven room house. 2300 SF of floor. Paint all walls, ceilings and trim in bathrooms, kitchen, bedrooms, living room and family room.

Long Range Plan

Total time on job: five days — 120 manhours. Prepare and paint the bedrooms, living room and family room in four days. Kitchen and baths to be completed in one day.

Daily Plan

First day: Do all prep work in bedrooms and family room.

Second day: Spray acoustic ceilings in bedrooms and family room.

Third day: Paint walls and do final sanding of trim in bedrooms and family room.

Fourth day: Paint trim in bedrooms and family room. Prepare baths and kitchen with undercoat.

Fifth day: Paint bathrooms and kitchen.

The foreman doesn't have to see the job to make plans like these. If he can visit the job the day before work starts, fine. But it's not essential. He has all the information needed in the Job Work Order and the estimate.

The work can go as planned, or it may be changed as needed because the carpenter wasn't quite finished in the kitchen. The point is that you have a plan to follow, not that you stick to the plan precisely.

Production Targets

Every crew leader should have a production target for his crew. Meeting production targets is the key to winning in the painting game. The original target is set by the estimator. If the estimated and actual manhours required are nearly the same, both the estimator and the foreman have done their jobs. If there's a wide difference between estimated and actual hours, both the estimator and the foreman have egg on their face.

From the original manhour target, the foreman sets individual targets for each painter for each day. The foreman checks and resets these targets after the morning break, after lunch and again after the afternoon break. He always lets the painters know what the current target is.

Paint Contractor's Manual

Job Work Order

Job name: _____ Job number: _____

Job address: _____
_____

Phone: _____

Contact name: _____

Start date: _____ Estimated completion: _____

Work summary: _____
_____
_____
_____
_____

Special instructions: _____
_____
_____
_____

Total estimated manhours: _____

Job work order
Figure 6-1

To be effective, targets must be reasonable and reflect the ability of the painter and the difficulty of the work. The foreman should have at least the tacit approval of each painter that the goals are realistic.

Targets build predictability into your work. Predictability reduces risk and increases the opportunity for success.

It's the foreman's job to meet production targets. If a foreman's consistently below target, either the foreman or the estimator is not doing his job. It's time to adjust the manhour standards or improve the effectiveness of the foreman.

Job Completion Certificate

When the job is nearing completion, the foreman should review the job estimate to be sure that all the estimated work has been done. There's nothing more embarrassing than trying to collect from a client who can point to a door that wasn't painted as proposed.

Have the foreman fill out a Job Completion Certificate as shown in Figure 6-2 before the crew leaves the site. It's the foreman's assurance that all work has been completed.

Whenever possible, get the client's signature on the Job Completion Certificate before the crew leaves the site. This confirms that the work has been done as agreed. Don't ask for the client's signature days later when he's had hours to find every minor holiday (missed spot), drip, and splatter. Get the signature while the crew is cleaning their brushes and before the paint dries. Signing the certificate is agreeing that payment is due on the job.

If some minor touch-up work remains to be done, that's O.K. The client can note that on the form. Get the client to list everything that he wants touched up. This is important. You're not promising to come back in several days and spend hours repainting everything the client can think of. You're just agreeing to touch up the items the client lists on the form right now. This list of discrepancies becomes the final list of work yet to be done. When that list is worked off, the job is complete. Naturally, the shorter the list, the more praise is due the foreman.

The Field Supervisor

When your company reaches a point where several jobs are going on at the same time, your main responsibilities become promotion, estimating, phone calls and paperwork. There may not be enough hours in the day to supervise field work too. When that happens, it's probably time to promote one of your most dependable and knowledgeable foremen to the job of field supervisor. Promoting someone to this position will nearly always be better than hiring an experienced supervisor. Of course, the new supervisor has to know how to handle painting crews. But his most important qualification is knowing exactly how *your* company works.

The primary responsibility of the field supervisor is to see that all jobs are completed in the most efficient and professional manner. Other responsibilities include:

1) Assigning crews to handle each job

2) Getting colors and samples approved

3) Making sure that each foreman has the materials and equipment needed

4) Checking each job daily

5) Making the final check on each job as it's completed

For a field supervisor who's doing his job right, a typical day might go like this:

Before leaving the house, he'll make phone calls to adjust crew assignments for the day. Having assigned painters to cover each job, he leaves home with a list of last-minute requests for materials and equipment needed that day. His first stop will be at the paint store to pick up what the foremen need. Then it's time to visit all the job sites. He checks with each foreman about job progress, problems, and production targets. He also checks on the quality of work, makes sure that company procedures are being followed, and assures himself that all materials and equipment needed to complete the day's work are on hand.

The remainder of the day will be spent at the office or meeting with customers and contractors. Office work includes discussing job progress with the contractor and taking care of personnel matters such as payroll, hiring, timekeeping, and scheduling.

Job Completion Certificate

Job name: _____ **Job number:** _____

Address: _____

Start date: _____ **Finish date:** _____

Evaluation by Foreman

Completion checklist:	Yes	No	Comment
All doors & windows rehung?	☐	☐	_____
All door hardware replaced?	☐	☐	_____
All window hardware replaced?	☐	☐	_____
All latches & handles replaced?	☐	☐	_____
All windows cleaned of excess paint?	☐	☐	_____
Touch-up paint left with client?	☐	☐	_____
Rags and empty cans disposed of?	☐	☐	_____
All tarps & masking tape removed?	☐	☐	_____
All furniture & furnishings replaced?	☐	☐	_____
Overspray removed from all surfaces?	☐	☐	_____

I certify that I have reviewed the agreement for this job and have determined that all surfaces have been prepared for painting and then painted in compliance with that agreement, including modifications.

Signature of Foreman _____ Date _____

Evaluation by Our Customer

Please complete this evaluation before the foreman and crew that worked on your project leave the site.

	Yes	No	Comment
Was work done as agreed?	☐	☐	_____
Has all trash been removed?	☐	☐	_____
Was all work neat and professional?	☐	☐	_____

I have checked the work that was done and find it completed to my satisfaction with the exception of the items noted below.

Exceptions _____

_____

Signature of Client _____ Date _____

Job completion certificate
Figure 6-2

Planning the Job

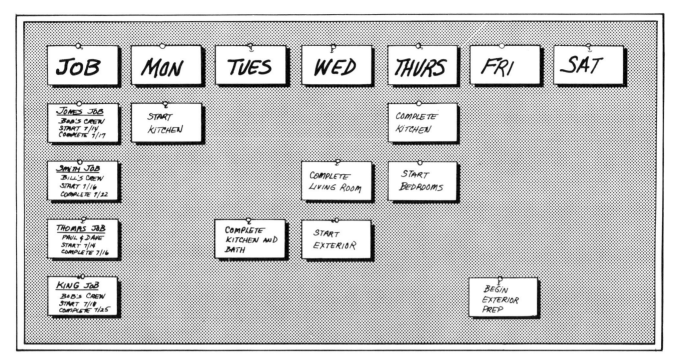

Job scheduling board
Figure 6-3

Job Scheduling Board

Even small painting companies sometimes have two or more jobs going at the same time. There may be a new job to start, one in progress, and a completed job that needs a little touch-up.

I've found that a job scheduling board makes scheduling easier. The board should show jobs in progress this week, and jobs to be done in the next two weeks. Information on each job should include the crew assigned, the scheduled or actual start date, the stage of completion if work has started, and the anticipated completion date.

I use a 3' x 8' cork bulletin board for job scheduling. A big sheet of posterboard will work nearly as well. Write the information about each job on a 3" x 5" index card. Pin these cards up at the left side of the board. See Figure 6-3. Write the names of the days of the week on other 3" x 5" cards and pin these up across the top of the board. You'll probably need to schedule three weeks in advance, so have three sets of cards with the days of the week across the top of the board. (Figure 6-3 shows only one week, but your board needs at least three weeks.)

Under the days of the week, and opposite the name of each job, pin up cards that note key milestones in each job. You don't need a card posted under every day that a job will be active, but be sure to post a card that notes the start day, and another under the day when work should be completed.

It's easy to move these cards around the board as the status of each job changes. That's important because your scheduling board is a waste of time if it isn't up-to-date. If work is done on schedule, no change is needed. But any change in the schedule will require a change in the board.

On Monday morning, all milestone cards for the week just past are taken down and all the remaining cards are moved one week to the left. That makes room to post milestone cards for work to be done three weeks in the future.

The scheduling board serves several purposes. First, any foreman, painter or client can see at a glance what jobs are active, what work has been done this week, and what work is coming up in the next two weeks. Second, if two or more people are scheduling jobs or advising clients when work is to start and be completed, the board becomes the collection point for all schedule information. A client who calls in to request information about the start date or finish date will always get an answer, regardless of who takes that call or who's managing that job. Third, the board makes it easy to spot conflicts, anticipate material and equipment requirements and make adjustments when a delay in one job will affect start dates on other jobs.

There's one more good reason to have a scheduling board. Everyone in my company watches the board. Everyone knows who's assigned to each job. The foreman's name is always posted under the name of the job. No foreman wants to see his job slip behind schedule. The painters know the schedule board back at the office says exterior work is supposed to start today. If it doesn't start today, the milestone card has to be moved one day to the right. If exterior work can be started one day early, the milestone card is advanced one day. And everyone notices it.

That's why I feel the biggest value to the scheduling board is not in scheduling. It's the recognition it gives for good work and the penalty it imposes for substandard work. The key steps in the job take on extra importance when foremen and painters know everyone's watching job progress. The board gives an incentive to improve production and reduce slippage. It gives every painter a feeling of participation in company goals. That's an advantage in any company.

Satisfying the Client

The most important job you do isn't painting. It's giving your client the service he expects. Scheduling carefully, and sticking to that schedule, promotes customer satisfaction. Making the work go quickly, and with as little disruption as possible, helps too. But many clients want more than quality work and efficiency. Many expect professional advice on color selection.

Selecting Colors

Although it isn't your job to select colors, many clients will ask your opinion before choosing colors. Here are a few of the basics every painter should know about color.

Color or combinations of colors set the mood of a room. There are right colors and there are wrong colors for every situation. Since the labor and material cost will be about the same regardless of the color selected, it's worth the time and trouble to find the best color combination.

No paint contractor can get very far without some paint samples. Most paint stores will supply all the color swatches you need. Show your client how these samples match the furnishings and how they look in the room you're painting.

All the furnishings in the room, from the carpeting to the drapes to the furniture itself, reflect color and light. This reflected color alters the way all other colors in the room are perceived. A dominant furniture color, particularly if it is a deep, rich tone, will be reflected all around the room and influence how all other colors in the room appear.

The intensity of light in most rooms varies during the day. Light intensity affects tone and color. Still another tone will be created at night by artificial lighting.

The only sure way to see how a color is going to look is to buy a quart and paint a large sample on the wall. Give your client several days to decide so he can see how the sample looks at different times of the day and night.

Most painters will agree that painting the ceiling, walls and trim the same color seems to reduce the ceiling height and make the room smaller. Monochrome color schemes close the room in. Painting the ceiling off-white and the walls a different color raises the ceiling and visually enlarges the room.

Coordinate the trim color with the furnishings and accessories to tie the furnishings or accessories to the room.

Pure white is too stark for most people's taste. A soft off-white seems "white" once the room is painted. Most paint stores have color charts that show which colors go well with each other.

The final choice of color should always be made by the client, decorator or architect. Be ready to advise, but let others select the final color. The most important rule is to avoid surprises. Never apply a color until it's been approved by your client. If color will be critical, apply several square feet and then get your client's comments before going ahead.

If the client insists that whatever color you choose will be fine, note that on your bid or contract. Otherwise, have the agreement state what colors are to be applied and the name of the manufacturer.

Color Schedules

A color schedule is simply a list of which colors go where. You'll usually get a color schedule for a

Planning the Job

residential job when a decorator or architect is involved. This makes it easier for you. No suggestions are needed.

Figure 6-4 shows a good color schedule prepared by a decorator. Your job is to follow the schedule exactly. Consult with the client and the decorator if you have any questions or expect any problems with paint selections. Look it over carefully in advance. A common problem is that a color for some part of the job is left out. Then you have a crew waiting while the foreman tracks down the decorator to find out what color goes on the trim in the dining room.

Tips on Buying Colors

Coverage varies with color. You would expect that dark green or blue or brown would cover better than white. They don't. The very dark, deep colors are less likely to hide the paint being covered than a white or off-white. You can figure on a minimum of two coats when using deep colors. Sometimes as many as three or four coats are needed.

It's the paint base that determines hiding ability in paint. Dark colors use a clear base. Whites use an opaque base. Opaque base won't work with darker colors. The base can't absorb enough tint to take on a deep, rich color. Beyond a certain point, the tint won't mix with the base. It simply falls to the bottom. Therefore, when mixing deep colors, paint manufacturers have to use a clear base that adds no color of its own. With this clear base, all hiding ability has to come from the tint. The base adds none. That's why light colors give better coverage. Both the tint and the base help hide the surface being covered.

One way to save time and money when painting with a deep color is to tint the undercoat as close as possible to the color of the finish coat. You won't be able to get an exact match. The undercoat uses an opaque base that will accept only so much tint. But getting close may be enough.

Tinting the undercoat will cut material cost quite a bit if you're painting an entire room or a lot of trim. It may also save the labor of another coat. Deep colors are more expensive because more tint has to be added. For the first coat, use undercoat at $10.00 a gallon rather than finish coat at $20.00 a gallon.

You can also save money on custom-mixed colors this way. If it looks like you will need two or three coats to hide the surface below, use a tinted undercoat for your first coat, not several gallons of custom-mixed paint. This works with both flat and enamel paint. Use as little custom-mixed finish paint as possible. It's expensive and won't be returnable. Try to buy only what you need.

Here's another tip. Most paint contractors buy paint by the gallon unless they're sure that only a quart is needed. Usually a gallon costs no more than three quarts. So, if it looks close, buy the gallon. You can always use the extra for touch-up or undercoating. And if possible, leave some paint for the client so he can touch up any nicks and scrapes that happen later. The client always appreciates this, and will remember it when he, or someone he knows, needs a painter.

Touch-up

It's a rare job that doesn't require some touch-up work. Always plan to spend at least part of an hour doing whatever touch-up is necessary to satisfy the client. The bigger the job, the more touch-up will be needed.

Some clients are more "picky" than others. Accept that fact. It's just part of the business. On a job where you're painting 8000 square feet of wall, 15 doors and jambs, 22 windows, 50 feet of outside handrail and the back porch swing, it's inevitable that there will be a few drips and a door jamb that isn't quite as smooth as the rest.

The best policy is to make touch-up a routine part of the job. When painting is finished, ask your client if he's noticed any spots that need touch-up. Do what's required and then tell the client that the work's finished. Most clients only want you to correct minor mistakes. Touch-up will take an hour or two at the most. And remember this. The touch-up list will be very short indeed if you did a good job in the first place.

Occasionally you'll get a nitpicker who *wants* to find mistakes. He'll crawl around on his hands and knees with a flashlight looking for a missed spot on the back of a closet. A client like this makes both the job *and* collecting the money unpleasant.

Getting into an argument with this type of client is usually a dead-end road. He holds the checkbook. Just do the work quickly with no protest. Of course, there's no rule that says you must be all smiles and sweetness. Just complete the touch-up in a professional manner, even though you know most of the items on the list are ridiculous.

COLOR SCHEDULE

140 South Street
Bensonville

DAY ROOM - LIVING ROOM - DINING ROOM

Walls: Dunn-Edwards Mojave Sand
Baseboard & Trim: Dunn-Edwards "White-White" satin
           or flat enamel
Ceiling molding: Mojave Sand

AREA BETWEEN LIVING ROOM & DEN

Walls: Mojave Sand
Baseboard: "White-White" enamel
Cellar door: Mojave Sand
Powder room: Pearl White
Stairway walls: Mojave Sand
Stairway trim: "White-White" enamel

UPSTAIRS LANDING & HALLWAY

Walls: Mojave Sand
Linen cabinet & closet door: Mojave Sand satin enamel
Doors to master bedroom, children's bedrooms & bathroom:
     "White-White" enamel

BOY'S BEDROOM (SOUTHWEST CORNER)

Walls: To be wallpapered
Ceiling: "White-White"
Baseboard: "White-White" enamel
Closet & closet door: "White-White" (walls flat, door
                              enamel)
Trim: "White-White" enamel

Typical color schedule
Figure 6-4

Of course, there may come a time when it gets absurd. If you're on your third day of touch-up and have painted the trim four times, it may be time to demand your money and *threaten* legal action. But find some way of resolving the dispute if possible. Unless your client is totally unreasonable, almost any settlement is better than going to court.

Say you've just invested three weeks on a job and have advanced $1,500 for materials. That's a bad position to be in. Your client should have been making progress payments each week. Maybe he's holding out now because he knows you're over a barrel. Anyhow, he's now holding up an $8,000 final payment. Is it better to sue or settle for less than $8,000? My advice is to threaten suit all you want. But don't sue, even if you have to settle for only $5,000 or $6,000.

If you use an attorney for collection, his fee will be a third to a half of the amount collected, depending on how quickly the money is collected. The longer it takes, the higher the fee. An attorney could send a letter to your reluctant client. The fee for that will probably be only $50. But your client will probably seek out legal advice of his own when he gets the letter.

Your client's lawyer has to make a living too. He'll advise his client to withhold payment, insisting that he can whittle the charges down enough to cover his fee. At that point you've got two lawyers taking pot shots at each other, one charging you directly and the other earning his fee by reducing what you collect. There's no percentage for you in that game!

My advice is to try every way to settle the dispute yourself. When a dispute ends up in two law offices, the real winners are the lawyers.

Plan for Safety

Accidents cost money. Plan to avoid them. Follow a few common-sense rules and you'll eliminate most of the common accidents. The foreman is always on the site. It's his job to check for violations of safety rules and make corrections on the spot.

Fumes

Many materials used in painting give off dangerous fumes — especially in an unventilated room. Exposure to dangerous fumes causes nausea and eventually intoxication if it lasts long enough and if the fumes are sufficiently intense. Intoxication from paint fumes is a standing joke among many painters. But there's nothing funny about it when you're sick from them.

Be alert for feelings of dizziness, light-headedness or disorientation. If it starts to happen, go outside and take a walk. Fresh air will solve the problem. After your head clears, take a few minutes to improve ventilation so fumes don't accumulate again.

If you're working in close quarters, keep windows and doors open. Use a fan to draw fumes out of the room. When spraying lacquer, stain or enamel, always wear a safety mask.

When you're working in a home that's occupied, make sure the occupants know about the possible danger. Ask them to keep children and pets out of the area.

Ladders

Ladders are a necessary part of the painting business. But misuse of ladders causes most serious painting accidents. Buy good ladders. Give them proper care. Discard ladders that are unsafe. Make sure every painter on your payroll follows these rules:

1) No matter how tempting it is, *don't stretch too far* trying to reach that last six inches. Stretching *may* save the couple of minutes it takes to move the ladder, but at the risk of a serious (and expensive) injury. It's not worth it. If you find you'll have to stretch, climb down and move the ladder. Figures 6-5 and 6-6 show the wrong and right ways to paint from the top of a ladder.

2) Make sure that the foot pads are securely set into the ground before climbing the ladder.

3) When possible, have someone support the ladder during ascent and descent.

4) Don't even come close to electric wires when working outside.

5) Never walk away from a stepladder with a paint bucket sitting on the ladder shelf.

6) Don't be cheap with ladders. If you don't want to buy good ladders, rent what you need. You certainly can't afford the accident that unsafe equipment will cause.

Paint Contractor's Manual

**Stretching too far on a ladder
Figure 6-5**

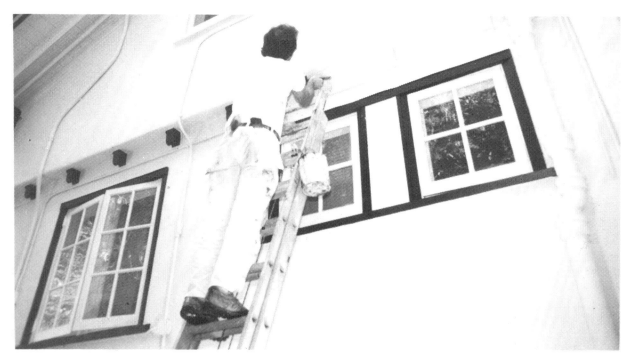

**Proper work position on a ladder
Figure 6-6**

Planning the Job

7) Make sure the ladder is placed at the right angle. Figure 6-7 shows the proper angle for an extension ladder. Figures 6-8 and 6-9 show the ladder placed at dangerous angles.

Other Safety Tips
In general, a clean and well-organized work area is a safe work area. Encourage your crews to clean the job site as they work. That cuts down on accidents. Here are some specific rules to follow:

1) All used rags should be picked up at the end of the day and disposed of in a metal container. Never leave used rags on the job.

2) Collect all tools at the end of the day and put them in a tool box.

3) Dispose of any used razor blades immediately. Either slip them into the blade disposal section of the blade dispenser or wrap masking tape around the entire blade several times.

4) Dispose of used thinner properly or keep it in a tightly-closed metal container. Be especially careful of this if there are small children in the house.

5) Keep all paint in closed containers when not in use.

It's just good professional practice to be safety conscious. A painting crew that has several accidents will become the laughingstock of the neighborhood where they're working. That's not the kind of publicity you need.

Plan for the Right Equipment
To get top productivity from all painters, make sure they have the right tools. There are two categories of tools: those you supply and those the painters are expected to supply themselves.

Basic Tools Supplied by the Painter
There are several reasons why painters should furnish some tools themselves. First, painters who supply their own small tools are more likely to have them when needed. Second, people take better care of their own equipment than equipment loaned to them for their use. If you provide the brushes, you'll spend a lot of money replacing brushes. It's easy to forget to clean a brush if the boss will buy a new one.

Another reason why painters should furnish the basic painting tools is efficiency. A change in the schedule may require switching painters from one job to another. If each painter has his basic tools with him at all times, he's always ready to move to another job. And when he gets there he's ready to go to work.

Painters on your payroll should be required to furnish *at least* the following tools. Each painter is responsible for these tools and must replace them if they're lost or damaged.

- Two or three brushes of different sizes for water-based paints

- Two or three bristle brushes of different sizes for oil-base paints

- Duster (an old paint brush with the handle sawed off)

- Phillips-head and regular screwdrivers

- Two or three putty blades of different sizes

Tools Supplied by the Contractor
You should supply all major tools, disposable equipment and supplies. These include:

- Drop cloths

- Rags

- Ladders and scaffolding

- Spray equipment

- Paint and thinner

- Putty, sandpaper and spackle

The Small Shop
Consider outfitting each painter with the basic supplies needed on most jobs. If a painter has these supplies, he doesn't have to wait for any other supplies to begin work:

Paint Contractor's Manual

Ladder placed at proper angle
Figure 6-7

Ladder placed too far away
Figure 6-8

1) One gallon of thinner

2) One can of painter's putty

3) One can of spackle

4) A couple of cans of undercoater or sealer

5) Several grades of sandpaper

6) One enamel roller cover

7) One flat roller cover

8) One 6-foot roller pole and cage handle

9) One roller grid for a five-gallon bucket

10) One empty five-gallon bucket

11) Several one-gallon buckets

12) Two average-size drop cloths

13) Several rags

A painter who arrives with his own small tools and with these basic supplies is ready for work. I call this collection of basic tools and supplies a painter's *small shop*. The small shop fits easily in the trunk of a car and should be with each painter

Planning the Job

**Ladder placed too close to building
Figure 6-9**

at all times. Every painter should be responsible for maintaining his small shop, replacing his own tools and requesting replacement of supplies and disposable equipment.

Brushes

I can't leave the subject of equipment without recommending that you use good (and therefore more expensive) brushes. Although the initial investment is higher, good brushes last longer and do a better job. A good brush will cut in a clean straight line, hold more paint, provide an even flow and give better paint distribution. Even though my painters buy their own brushes, I insist that their brushes be of top quality.

I'll have more to say on brushes in the next two chapters, including choosing the right brush for the job, and how to take care of them.

Getting Ready to Paint

If you follow my recommendations in this chapter, you'll run more efficient jobs and avoid many of the common problems that plague paint contractors. The next chapter gets us a little closer to actual painting: selecting the paint, setting up the shop, and preparing the site for production.

149

Chapter 7

Preparing to Paint

Once you've planned the job, scheduled it, and assigned the foreman and crew, it's time to get ready to paint. Your goal is to prepare the surface and apply coatings as efficiently as possible while maintaining good professional standards. That's a formula that puts profits in your pocket and produces satisfied customers.

Preparation includes selecting the right tools and best paints for the job. Unless a decorator or architect has provided a color schedule and specified brand names, the choice of paint is probably up to you. Pick good quality paint that's easy to work with and will cover fully with one or two coats.

Paint Selection

Paint is either solvent-thinned or water-thinned. Solvent-thinned paints are most commonly oil-base paints, although some specialty coatings such as epoxies, polyesters, and urethanes which are not oil-base are also solvent-thinned. Enamels are made with a varnish or resin instead of the usual linseed oil vehicle, and fall in the oil paint group. The vehicle or base in oil-base paint usually consists of alkyd resin with turpentine or mineral spirits as the thinner.

Oil-base paints are very durable, are highly resistant to staining and damage, can withstand frequent scrubbings, and give good one-coat coverage.

Water-thinned paints are usually latex. But some non-latex paints are water-thinned. Latex paint has fine particles of resin emulsified (suspended) in water. Latex paint dries faster, usually holds its color better, resists alkali and blistering, and makes clean-up easier. All surfaces should be dry when applying paint. But some latex paints can be applied on slightly damp surfaces.

Paints for Exterior Surfaces

Figure 7-1 shows paint recommendations for exterior surfaces.

Preparing to Paint

PAINT CHOICE

SURFACE	Aluminum Paint	Asphalt Emulsion	Awning Paint	Cement Base Paint	House Paint (Oil)	House Paint (Latex)	Metal Primer	Porch-and-Deck Enamel	Primer or Undercoater	Roof Cement or Coating	Spar Varnish	Transparent Sealer	Trim-and-Trellis Paint	Water Repellent Preservative	Penetrating Wood Stain (Latex or Oil)
MASONRY															
Asbestos Cement					X●	X			X						
Brick	X			X	X●	X			X			X			
Cement and Cinder Block	X			X	X●	X			X			X			
Cement Porch Floor						X		X							
Stucco	X			X	X●	X			X			X			
METAL															
Aluminum Windows	X				X●	X●	X						X●		
Galvanized Surfaces	X●				X●	X●	X						X●		
Iron Surfaces	X●				X●	X●	X		X				X●		
Siding (Metal)	X●				X●	X●	X●						X●		
Steel Windows and Doors	X●				X●	X●	X						X●		
WOOD															
Frame Windows	X				X●	X●			X				X●		X
Natural Siding and Trim											X				X
Porch Floor								X							X
Shingle Roof														X	X
Shutters & Other Trim					X●	X●			X				X●		X
Siding					X●	X●			X						X
MISCELLANEOUS															
Canvas Awnings			X												
Coal Tar Felt Roof		X								X					

Note: X = Paint choice
● = Primer or sealer may be required. Check container label.

Exterior paint selection chart
Figure 7-1

Barn walls— Walls in farm buildings get rubbed by animals and have to stand up under frequent washings. Storage building walls also get hard use. Durable paint is required. But lead-base paint can't be used because it could poison the animals. Catalyzed enamel, epoxy, polyester and urethane paint are good choices on better-quality farm jobs. They're expensive, but they're also more durable and washable. These paints usually come in two containers and must be mixed before application. Follow the label instructions when using these types of paint.

Wood siding— Either latex or oil-base flat or enamel can be used. An oil-base primer is recommended over resinous woods (such as pine) and over wood that tends to bleed (such as redwood and western red cedar).

Trim, windows, shutters, and doors— Because wood trim is usually treated with a water-repellent preservative before finishing, any form of latex or oil-base paint or stain can be used. Latex exterior enamels dry fast, have good color and gloss retention, and are durable. Oil-base paint may not stay glossy as long. Avoid any paint that becomes chalky. It will discolor adjacent surfaces.

Masonry— Exterior latex masonry paint is the standard for masonry. Cement base paint may be used on nonglazed brick, stucco, cement, and cinder block.

Here's my recipe for an inexpensive, attractive masonry paint. Mix one part of hydrated lime with five parts white Portland cement. Add water until the mixture is about as thick as condensed milk. Add any good grade of mineral coloring to get the tint you want. Then add two parts of fine sand to the mix if you're painting a rough surface like cinder block. Excellent block fillers are also available in better paint stores.

Dampen the surface before applying the paint. Brush or spray the paint on. A short, stiff-bristled brush will help fill pores. A heavy nap roller also works. Don't let the paint dry too fast. It has to dry slowly for proper curing. After it's firm, wet it down with water mist several times during the next 48 hours.

If you want to paint over this mixture with some other paint, apply a sealer before repainting.

Galvanized iron— Ordinary house or trim paint is acceptable as the finish coat on gutters, downspouts, hardware and grilles. But be sure to use the right prime coat under the finish coat. For instance, a metallic zinc dust primer is recommended on galvanized metal. Red lead or zinc chromate primer is used on iron. Special types of enamel are available for window screens. Some types of metal require specific surface preparation techniques and specially formulated coatings.

Concrete or wood porches and steps— Porch and deck paint can be used on both cement and wood. First, treat the wood with a water-repellent preservative solution that's suitable for use under the paint you've chosen. Then apply an oil-base primer.

Paints for Interior Surfaces
Almost every new or bare surface requires a primer or prime coat before application of the top coat. The primer is designed to both protect the surface and provide good surface adhesion for the finish coat. In a pinch, the top-coat material can be used as a prime coat. Then a second and possibly third top coat can be applied.

Acoustical surfaces require special treatment. You don't want the paint to alter the acoustic properties of the material. That's why only a thin coat is used. This usually will have little effect on sound characteristics.

Figure 7-2 is an interior paint selection chart. Use it to select the right paint and primer for each type of surface listed. These primer/top coat combinations are good general recommendations. But the label on the can will provide more specific instructions.

The Right Tools
Choosing the right tools is the next step in preparing for the job. Whether you brush, roll or spray, the right equipment will make the painting go faster, with fewer mistakes to correct.

The Right Brush for the Job
We touched briefly on brushes in the last chapter. I recommend only good quality brushes. They last longer and do a better job. Once you've decided to invest in good quality brushes, the question is what type of brush to buy, and what size.

Preparing to Paint

PAINT CHOICE

SURFACE	Alkali Resistant Enamel	Alkyd Exterior Masonry Paint	Alkyd Flat Enamel	Alkyd Floor Enamel	Alkyd Glossy Enamel	Alkyd Semi-Glossy Enamel	Epoxy Enamel (Opaque)	Epoxy Finish (Clear)	Lacquer	Latex Exterior Masonry Paint	Latex Flat Wall Paint	Latex Floor Enamel	Latex Glossy Enamel	Latex Semi-Glossy Enamel	Pigmented Wiping Stain	Portland Cement Masonry Paint	Portland Cement Metal Paint	Shellac	Urethane Enamel (Opaque)	Urethane Finish (Clear)	Varnish
MASONRY																					
Brick	X 11	X 11	X 8,11		X 8,11	X 8,11	X 11,7			X 11	X 8,11		X 8,11	X 8,11					X 11,7		
Cement Block	X 11		X 4,7		X 11	X 4,7	X 4,7				X 4,7		X 4,7	X 4,7		X 11			X 11,7		
Ceramic Tile Flooring				X 11			X 11					X 11							X 11		
Concrete	X 11		X 4,11	X 11	X 4,11	X 4,11	X 11				X 4,11	X 11	X 4,11	X 4,11		X 11			X 11		
Concrete Flooring	X 11			X 11			X 11					X 11							X 11		
Drywall			X 6		X 6	X 6	X 6,11				X 6,11		X 6	X 6					X 6,11		
Plaster			X 6,2		X 6,2	X 6,2	X 6,11				X 6,11		X 6	X 6					X 6,11		
METAL																					
Aluminum			X 1		X 1	X 1	X 1				X 1		X 1	X 1					X 1		
Galvanized Steel			X 14		X 14	X 14	X 14				X 14		X 14	X 14		X 10	X 10		X 14		
Iron and Steel			X 1,5		X 1,5	X 1,5	X 1,11				X 1,5		X 1,5	X 1,5		X 10	X 10		X 14		
Steel Flooring				X 11			X 11									X 10	X 10		X 11		
WOOD																					
Flooring				X 11			X 11	X 11	X 11		X 11			X 13,12				X 11	X 11	X 11	X 13,12
Trim and Paneling			X 3		X 3	X 3	X 3,11	X 11	X 11		X 3		X 3	X 3	X 13,12			X 11	X 3,11	X 11	X 13,12
MISCELLANEOUS																					
*Accoustical Tile			X 2								X 11										
Vinyl Wallcovering, Smooth, with Design			X 11		X 11	X 11					X 11		X 11	X 11							
Vinyl Wallcovering, Smooth, without Design					X 11	X 11					X 11										
Vinyl Wallcovering, Textured			X 9,11		X 9,11	X 9,11					X 11										
Wallpaper			X 6,2		X 6,2	X 6,2					X 6,2		X 6,2	X 6,2							

CODE

1. Alkyd metal primer
2. Alkyd primer
3. Enamel undercoater
4. Exterior masonry paint
5. Latex metal primer
6. Latex primer
7. Masonry block filler
8. Masonry surface conditioner
9. Oil-base primer
10. Portland cement metal primer
11. Topcoat material used as primer
12. Wood filler
13. Wood sealer
14. Zinc-rich metal primer

Note: X = Paint choice
Numbers = Primer choice (code in column on right.)

Interior paint selection chart
Figure 7-2

For latex and water-base paints, use a nylon and polyester filament brush. It holds its shape in water-base paints and is easy to clean. The synthetic filament won't soften and become limp. A natural bristle brush would absorb water.

For oil-base paints, enamels, varnishes and lacquers, use a natural bristle brush. Natural bristle tends to regulate the flow of oil-based paint, leaving a smoother finish and requiring less dipping.

The Right Size Brush
Use the right size brush for best results and top production rates. A professional painting company should have a good selection of brush sizes on hand. Here are the standard sizes and their uses:

1" brush — Small trim, shutters, touch-up

1½" to 2" brush — Window sash, shutters, molding

2½" to 3" brush — Cabinets, baseboards, shelves, doors, beams, fences, gutters, stair steps, handrails

3½" to 4" brush — Walls, ceilings, floors, paneling, exterior facing

Care and Cleaning of Your Brushes
A brand new brush is the ideal — clean looking and smooth working. Brushes stay this way if you clean them carefully after each use. The handle will collect some paint, but this accumulation comes off easily with a wire brush and some paint brush cleaner.

Bristle brushes— should be cleaned as soon as possible when work is finished. Use the same solvent recommended for thinning the paint you're using. Pour about a pint of thinner into each of three clean one-gallon buckets. In the first bucket, dip the brush several times. Then comb and wire brush the bristles. Spin the brush with the handle held between both hands, or use a *spinner*. The spinner (shown in Figure 7-3, Step four) gives a greater spinning action, which helps eliminate any particles attached to the brush fibers. Always hold the brush inside an empty bucket when you spin. Failing to do so will provide interesting results on you, the entire room, and anyone else in the room.

Then dip the brush in the second bucket, rinse the bristles, and spin again. By the time you rinse your brush in the third bucket, it should be clean. Test for remaining paint residue by dipping it in the thinner and squeezing the bristles. The thinner that runs out should be clear. If not, repeat the procedure. Change the thinner frequently if you're cleaning several brushes. Figure 7-3 shows the cleaning procedure step by step.

The spinner is also useful for cleaning roller covers. Slip the cover off the clips at the end, and spin.

Use as many washes as needed. Be careful to work the solvent into the center of the brush and down to the handle. A brush comb will help remove paint residue and will straighten the bristles so they dry straight. Stick the teeth of the comb into the bristles right at the heel of the brush. Move the comb toward the end of the bristles. But pull the comb up and out about 1/4" before it reaches the end of the bristles to help prevent excessive wear on the ends. Try to avoid the tips as much as possible when using the wire brush.

Many painters use kerosene or kerosene with a little motor oil in it as a last rinse. This keeps the bristles soft. If you use this final rinse, be sure to rinse the brush in clear thinner before using it the next time.

Use strong or "hot" solvents only as a last resort. They take the life out of bristles and remove all oils. Soaking a brush in a water-type brush cleaner is never recommended. That destroys the bristle and ruins a good brush.

When almost dry, put the brush back in its brush cover. That reshapes the bristles so they look just like they did when new. This brush is now ready for the next job.

Nylon and polyester brushes— are usually used for fast-drying, water-thinned paints. But many painters find nylons harder to clean than bristle brushes. They require more washes and rinses.

When cleaning a synthetic brush, follow the same procedure as for natural brushes, but use water instead of paint thinner. Use soap or detergent if necessary. Rinse repeatedly with clear water while using the wire brush. In severe cases, use paint thinner and then more detergent and water washes. Run the wire brush across the handle to remove paint that builds up there.

Preparing to Paint

Step one: Dip brush into thinner several times.

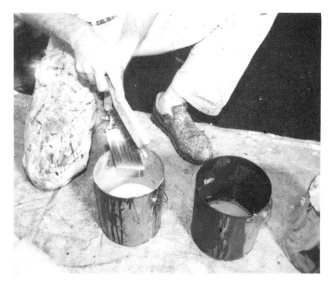

Step two: Use wire brush to clean the bristles.

Step three: Squeeze the bristles. If thinner runs clear, the brush is clean.

Step four: Spin the clean brush to remove excess thinner.

Cleaning enamel brushes
Figure 7-3

When a brush isn't cleaned carefully after each use, dried paint accumulates at the heel where the filaments join the handle. When that happens, the brush is said to be *heel hardened*. That results in *fingering* (the filaments form irregular groups instead of remaining in a uniform row), reduced flexibility, and uneven painting.

When a brush is heel hardened, rinse the nylon filaments with solvent until clean. Concentrate on the center of the brush, working solvent into the base of the heel. A brush comb always helps. But combing is particularly effective for getting to filaments at the center of the heel. Use the comb to remove paint residue and straighten the filaments.

Remember, a brush that looks clean may have inside filaments that are still saturated with paint, and crooked. Any paint left will harden, giving a permanent *set* to the matted bristles. Once the paint has set, no amount of cleaning or combing will restore the original brush shape and flexibility.

When using fast-drying water-base paint, work some water into the heel before using the brush. That makes cleaning much easier.

If you have a good brush that's taken a permanent set from hardened paint, it's probably impossible to restore the original shape. But you can remove most of the set. First, give it a thorough cleaning. Then dip it in boiling water. Finally, comb the filaments straight and hang the brush up until it's dry.

Smart painters waste very little money on brushes. They buy the best quality brushes available and then take exceptionally good care of them.

Brush Complaints — Causes and Cures
Soaking brushes in water will crack a wood handle. Strong cleaners take the life out of the bristles. Brushes should *not* be left to soak. The only thing to do with a brush at the end of a day is to clean it with the recommended cleaner. That keeps a brush as good as new.

Twisted nylon— Poor cleaning will cause a brush to finger and twist. A brush that isn't cleaned after each use will always do this. Remedy: clean well, straighten the filaments with a brush comb, and let the nylon dry completely. Warm water, detergent, and a brush comb can straighten most nylon brushes.

Soft and floppy bristle brushes— Cleaning with hot solvents can take the life out of a natural bristle brush. It's better to clean in milder solvents. Sometimes you can restore a "dead" brush by cleaning in kerosene or thinner with oil in it. Then let the brush stand shaped in the brush cover for several days.

Excess rounding off on the corners— This is caused by excessive use of the brush edge. Brushes are designed to be used flat, along the entire width of the brush. Use a smaller brush for those hard-to-get areas.

Hard-to-clean brushes— Many water thinned paints will dry completely in the brush during an 8-hour day. You'll have to use a strong solvent to clean the brush thoroughly. If this is happening, I suggest cleaning once *during* the day or at lunch time. Some painters make it a habit to change brushes at least once during the day.

Fingering— This usually happens after the brush has been used regularly for several weeks. The most common cause is improper cleaning. But sometimes a 6½" brush used on its corners or to paint a 4" surface will finger. Use the right brush for narrow surfaces. For a fingered brush, clean properly, comb well, and let the bristles dry, hanging straight. In most cases, the brush shape will improve in time for the next use.

Excessive side wear— This happens when a painter paints with the side of the brush instead of the tips of the bristles. Using the side of the brush causes unusual wear. The remedy is proper brush technique.

Excessive wear on the ends— When bristle brushes are used on rough surfaces, it's common for bristles to wear away or break. Like a scrub brush or a broom, paint brushes *do* wear out. But good painting technique with a quality brush keeps wear to a minimum.

Rollers
Polyester and nylon rollers are the most common. I recommend a synthetic roller with a 3/4" nap for smooth interior and exterior surfaces. A longer nap, 1" to 1½", works best on rough surfaces.

Lambswool rollers are used in production painting. They hold more paint. That gives more surface coverage between dippings. The disadvantage of lambswool is that it leaves a more textured surface unless it's rolled out (or back-rolled) carefully. When applying colors other than white, lambswool doesn't cover as well as a synthetic roller. The deep nap spreads the paint unevenly. A synthetic roller leaves a tight, even flow of paint on the wall.

There are advantages to lambswool rollers. As mentioned, they hold much more paint than synthetic rollers. They leave less overspray when rolling and are easier to clean than synthetics.

Lambswool usually costs at least twice as much

as synthetic rollers. But they should last twice as long if properly cared for.

The Purdy Brush Co. makes a polyester roller (Whitewing) which I've used and recommend. It performs and cleans up like a lambswool roller.

Whatever your choice of rollers, you're wasting time if you use anything smaller than a 3/4" nap with flat paint.

To clean water-base paint from a roller, rinse well with water and spin several times in a bucket.

Ladders
If you can have only one ladder, make it a four-foot wood painters' ladder. Don't buy a cheapie. Cheap ladders don't hold up. If the steps don't break, the bucket platform will. A cheap ladder is a poor investment and will become unsafe under constant use.

Four-foot ladders are enough for most interior jobs and single-story exteriors. Rent or buy other ladders as needed.

Buckets
One of the most frustrating tasks in painting is searching for a clean bucket. Plan ahead to have plenty of buckets ready to go. You'll never have too many.

The most common buckets are "fivers" (five-gallon buckets) and one-gallon cans. Use a "fiver" with a roller grid for rolling paint. Use a one-gallon can for trim work and cleaning brushes.

To get the most use out of your one-gallon buckets, use a can opener to remove the inside lip of the can. This provides more room for dipping brushes.

When you're finished using a bucket, remove all the paint by cleaning with solvent. Then turn the bucket upside-down with one edge propped up on a board or roller handle so it can air-dry. If you always do this, you'll always have a clean bucket available.

Setting Up the Shop Area
Once the paint and tools are ready to go, it's time to move to the job site and set up a shop area. This is the first task on-site. Even on the smallest jobs, set aside a shop area away from traffic. Find a protected area likely to be free of children, dogs, and neighbors. A garage, work room or enclosed patio is best.

Check the area for anything that might be damaged as painting progresses. Remove any bicycles or plants. Then spread out a plastic tarp. Cover it with a drop cloth, old sheet, or discarded carpet, drapes, or blanket. The plastic sheet acts as a membrane to keep spills from soaking through to the floor. The fabric placed on the plastic helps absorb spills so they don't get tracked out of the shop area.

If the shop area is set up outside or near an open door, it's a good idea to cover the corners of the tarp with bricks or paint cans. Otherwise, wind could get under the tarp, possibly spilling paint.

It's also a good idea to spread a drop cloth walkway for the first few steps from the shop area to the work area. This cleans accumulated paint and debris off your shoes as you leave the shop area. But always check your shoes before walking across carpet or tile.

The shop is where you store supplies, mix paint and clean brushes. This is dirty work. But it doesn't follow that the shop has to be a jumble of accumulated trash, tools, supplies and discards. Keep the shop as neat and orderly as possible. Encourage your painters to straighten up the shop and pick up around the job at least once a day. This is especially important when doing work in a home. A tidy painter presents a professional image that clients appreciate. It's especially important to clean up at the end of the day. This eliminates lost time looking for tools and materials and allows a running start the following day.

At the end of the work day, always cover the shop with another drop cloth. This keeps what's there out of sight and less of an attraction. If the shop is outside, lay down a sheet of plastic first. This keeps moisture out.

Figure 7-4 shows the shop on a residential job.

Getting the Room Ready
After the shop is set up, the next step is to move into the room to be painted and begin preparation. First, remove everything that can be moved easily (if your client hasn't done this already). If you're painting a house, empty out one or two rooms at a time. When those rooms are *completely* painted, move the furniture back and start on one or two more rooms.

Any furniture left in the room should be moved to the center and covered with plastic or a wax-

Paint Contractor's Manual

Setting up the shop area
Figure 7-4

coated disposable drop cloth. Disposable drop cloths are very inexpensive and can prevent damage to valuable furnishings. Old bed sheets don't make good drop cloths. Every drop of paint will soak through to the furniture.

Do whatever is necessary to provide adequate working space in the room. Paint is a liquid. Like every liquid, it spills and splashes. The more crowded the area and the more limited the floor space, the greater the chance of a damaging spill.

It took me several jobs to learn that a messy, cluttered work area usually results in a messy, unprofessional job. And the opposite is true. A neat, clean, organized work space usually makes for a high-quality, professional paint job. I offer this advice with the hope that you'll learn faster than I did.

Tools and Materials for Preparation

A 2" flexible spackle blade (putty knife)— This is the best tool for patching most cracks and holes. For large holes, use a wider blade.

Spackle— This is the most common patching material. It dries quickly and sands easily. Spackle comes pre-mixed and is very easy to work with.

Caulking gun and latex caulk— Use a gun to patch cracks where two walls meet, where the wall meets the ceiling or where the wall meets the door jamb, window, trim or baseboard. Latex caulk dries in minutes and can be painted without being primed.

Rags— The best rags are lint-free and absorbent. But any rag will do. Many paint stores sell rags by the pound. Use a damp rag to wipe excess caulk from the wall. That's the way to do a good-looking, clean, professional job.

Patching material for large or deep holes— Use a quick-drying patching material to fill in most of the hole. Leave the patch slightly underfilled. This type of material is difficult to work with and very hard to sand smooth. When the quick-dry patching material has set up, use spackle to bring the patch up flush to the wall.

Sandpaper— Use fine-grade sandpaper (number 220) for sanding spackle. Rougher sandpaper will leave scratches that are visible under the finish coat. Use a rougher grade (number 100 or 150) when sanding off rough edges around holes, cracks and peeling paint. Then patch and sand to finish the job.

Washing compound (T.S.P.)— Use T.S.P. to clean grease and heavy tobacco smoke off walls. T.S.P. (trisodium phosphate) comes in powder form and is mixed with water. Use rubber gloves if you expect to work with T.S.P. for very long. When you've finished washing walls and ceiling with T.S.P., rinse the surface thoroughly with a sponge mop. T.S.P. leaves a residue that prevents paint from bonding to the surface.

Do a Good Prep Job

The rest of this chapter is devoted to surface preparation. We'll cover interior preparation first and then go on to exterior work. Let me emphasize first that good surface prep is essential to *every* professional painting job. The most expensive paint, the best painting technique and the finest brush can't compensate for poor surface prep. Doing a good job of preparation actually saves painting time. Once you've learned good prep procedures, painting is a much easier task.

Painting is essentially the same from one job to the next. A door is a door and a wall is a wall. What changes is the preparation required. Every door requires custom preparation — a heavier sandpaper, a special primer, a different patching material. Preparation separates the men from the boys in the painting business. Learn professional prep methods and be sure your painters follow the recommended procedures.

Prepping the Room

Begin by making a quick survey of the work required. Use a broom and a spackle blade as you make this survey. While looking over the walls and ceiling, sweep the broom along the baseboards and above the door jambs and window casings. Then check for spider webs along the ceiling and in corners. Sweep them off. Use the spackle blade to probe peeling paint, deteriorated plaster, cracked wallboard and rotted trim.

Next, wash the walls if accumulated dirt is a problem. Any household detergent will work. For grease or a heavy buildup of cigarette smoke, use T.S.P. If you use T.S.P., be sure to rinse it off.

Wash off pencil marks, crayon scribbles and the like. If T.S.P. won't remove them, try a chemical cleanser like Formula 409. They probably won't come off completely. But once you get the loose material off, the surface can be primed with a stain-killing primer.

Patching cracks in wallboard and plaster— Figure 7-5 shows the procedure. Before you begin scraping and patching, cover the spot on the floor under the wall or ceiling where you're working. Using your spackle blade or a screwdriver, dig out each crack a fraction of an inch on all sides. Be sure to remove all loose material along the edges. Anything that's loose in the crack will come out very nicely, all by itself, after painting is finished. That ruins your paint job.

Next, run your spackle blade or sandpaper down the length of the crack. That should remove any protruding chips of old paint or plaster. Dust out the crack with a dry brush or blow the dust out — with your eyes closed.

Then wet the crack with a paint brush dipped in water. While the crack is still damp, apply spackle with your blade. Do this in one long stroke if possible. Two thin patch coats are always better than one thick one. Spackle that's too thick is hard to keep flush with the wall and tends to crack and shrink as it dries. If you are applying two coats, don't worry if the first coat isn't perfectly smooth.

When the first coat has dried, apply a second coat. Again, don't try to make it perfectly smooth. But be sure you have the crack filled at least flush with the wall. When the second coat has dried, sand with number 220 grade sandpaper. This will provide a nice, smooth finish.

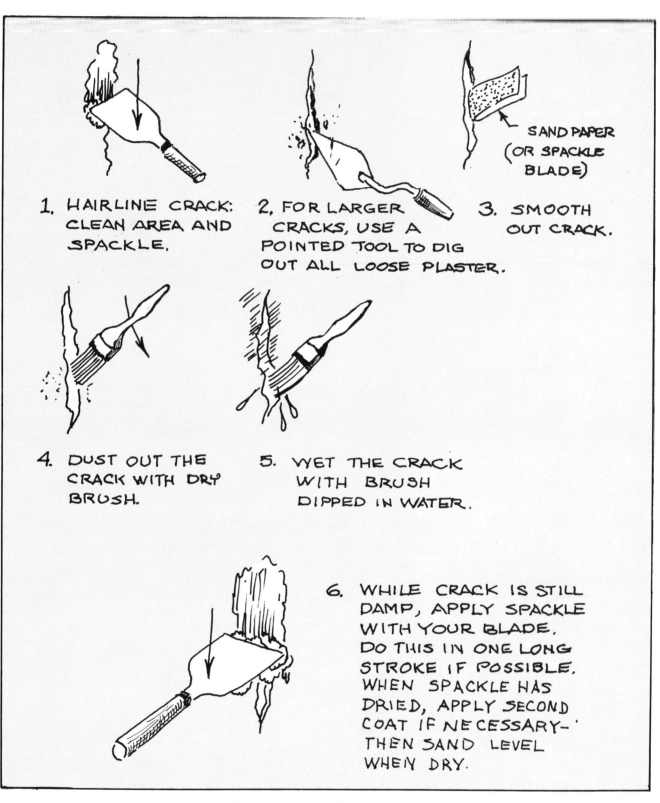

Patching cracks in wallboard or plaster
Figure 7-5

Patching holes in wallboard— Fill small holes with two applications of spackle, and sand smooth.

Holes up to two inches wide that go all the way through the wallboard need special attention. Here's the procedure. Remove any loose material around the edge of the hole. Stuff the hole with paper so there's something behind the hole to hold the fresh plaster. Then apply a layer of patching plaster. See Figure 7-6. Patching plaster comes in powder form to be mixed with water. It dries very quickly. That makes it handy for patching large, deep holes and cracks. When mixing, remember the thicker you mix it, the faster it dries. But don't make it so thick that it hardens before you can get it in the hole.

Leave the patching plaster slightly recessed, not flush with the wall. When the plaster has set up, apply a layer or two of spackle, which is much easier to sand. Sand until the patch is flush with the wall.

Here's the procedure for wallboard holes between two and six inches wide. First, remove any loose material around the edge of the hole. Cut a piece of stiff wire screen (old window screen works fine) slightly larger than the hole. Run a string through the center of the screen. See Figure 7-7. Wet the edges of the hole. Apply a coat of patching plaster all the way around the perimeter of the hole. Make sure some plaster gets on the back side of the drywall. Push the screen into the hole and let it expand so it laps over all edges of the hole from the inside. Use the string to pull the screen up tight against the backside of the hole and into the wet patching plaster. Place a brace like a pencil or short dowel across the hole. Tie the string to the brace so the screen is held tight in position as the plaster dries.

When the perimeter plaster has started to set, fill the hole with more plaster until the patch is almost level with the surface of the wall. When this plaster has set, cut the string and remove the brace. Wet the plaster around the area where the brace was and plaster in this spot. When all the plaster has dried, apply a layer or two of spackle with a wide spackle blade. Sand smooth to the level of the surrounding wall.

For wallboard holes larger than about 6 inches, it may be easier to replace the entire sheet of damaged wallboard.

Patching holes in plaster— Here's how to patch large holes in plaster: Again, dig out all the loose plaster from the edges of the hole. With your blade or a screwdriver, dig the edges of the hole back at a slant so the hole is slightly larger on the exterior surface than it is on the interior surface. This expands the contact area so the patching plaster bonds securely to the old plaster.

Then dust out the hole and wet it with water. Apply a layer of patching plaster with a wide blade. The depth of the plaster shouldn't exceed 3/8". If the hole is deeper than this, make two applications. While the plaster is still soft, etch a gridwork of lines over the entire surface with your blade. When this has hardened, wet the plaster and add another layer of patching plaster if necessary. If not, apply a coat or two of spackle and then sand smooth when dry.

If the hole goes entirely through the wall and the

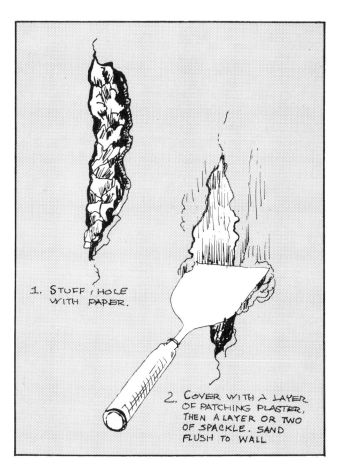

Patching small holes in wallboard
Figure 7-6

Paint Contractor's Manual

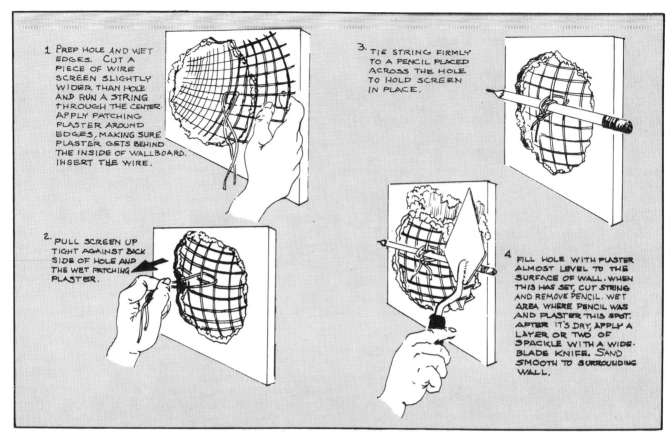

Patching holes up to 6" wide in wallboard
Figure 7-7

lath has been knocked out, use the screen method described in Figure 7-7.

Patch small holes in plaster the same way. Dig the edges of the hole back as described above. But patching plaster isn't needed here. Spackle will do the whole job.

Texturing patches— If the existing wall has something other than a smooth texture, your patch should match that texture as nearly as possible. Create a stippled effect by dabbing the surface with a sponge or stiff bristle brush while the patching material is still tacky. See Figure 7-8. Create swirls with a spackle blade. Dip the blade in water so it floats easily on the surface. Move the flat edge of the blade across the surface with a circular motion, the same way the plasterer's trowel was moved to create a texture in the existing wall. To match a rough-textured surface, mix sand in the final layer of patching material. It's possible to match nearly any surface. The only limit is your creativity.

Loose Paint

When cleaning and patching are complete, focus on the condition of the existing paint. Scrape away loose paint with a wire brush, spackle blade or circular power sander. Then sand the rough edges where the paint broke off. If the surface is still rough, apply a coat of primer where the paint was removed. If the edges are very prominent, lay down a thin coat of spackle and sand smooth.

Don't even consider repainting as long as there's loose paint on the surface. Painting over loose paint does neither you nor your client a favor. Today's high-pressure water cleaning equipment and modern power tools make it relatively easy to remove all loose paint. We'll discuss water blasting and rotary sanders later in this chapter when we cover exterior prep.

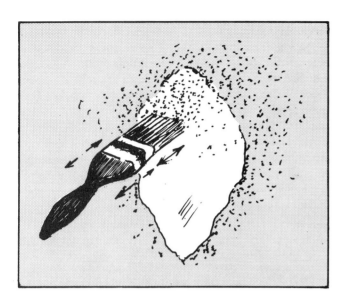

Texturing a patch with a stiff-bristled brush
Figure 7-8

Caulking Cracks

Now's the time to fill all the cracks. This isn't just a matter of aesthetics, of course. Sealing cracks cuts down on air infiltration and reduces heating and cooling costs. Some careful work now with a caulking gun will cut your client's fuel bills for many years.

Use latex caulk and a caulking gun to seal cracks around door jambs, window casings, baseboards, where walls meet, and where the ceiling meets a wall. Dig out the crack with your spackle blade or a screwdriver. When all loose material is removed, begin caulking. Hold the caulking gun at an angle and slowly run a line of caulk the length of the crack. Then run your finger over the line of caulk to even it out. Add more caulk where needed. Wide and deep cracks may require more than one application. Using a damp rag, wipe off any excess. This is very important. Dried caulk doesn't sand smooth. Sanding makes it roll up like a ball of rubber.

Covering Up for Painting

The furniture should be moved to the center of the room and covered with plastic or wax-coated paper drops. If you're painting the ceiling, cover the furniture and floor with a regular drop cloth laid *on top of* a plastic or wax-coated paper drop. Spreading out newspaper is no good. Newspaper shifts around as you walk on it, leaving part of the floor or carpet exposed.

If you're painting walls only, rip a paper drop in half. They usually have a seam down the middle to make separation easy. Place the halves against the wall. Even better, buy a 3' x 12' drop cloth called a *runner*. These are ideal when painting walls. A runner won't shift around like newspapers would. It's inexpensive and will last for years.

Cover light fixtures and doorknobs with paper or plastic held in place with tape. Here's how to mask around a light fixture. First, turn off the light. Loosen the screws that hold the base to the wall or ceiling. Pull the fixture away slightly. Don't twist the fixture at this point. Some old wiring may have deteriorated to the point where any movement can cause a short. Run a strip of tape around the fixture where it joins the wall or ceiling. Then cover the rest of the fixture, or remove the glass globe when you're painting in that vicinity. Turn the power to the fixture off if you cover it. Covered fixtures can overheat and shatter.

Be sure to cover items like air conditioners and curtain rods before beginning. Assume that everything in the room that could be splattered with paint will get splattered. Generally it's easier to move everything removable and cover everything that isn't. Worrying about paint drips slows production, and removing paint spots later is an unnecessary waste of time.

Preparing Specific Surfaces

Good surface preparation is an essential part of every painting job. Skimp on surface prep and you cut the useful life of a paint job by months or years. No matter what type of surface you're painting, it has to be clean, free of loose or checked paint, rust, scale, oil, grease, dirt, mildew and chemical residue. If the painter before you didn't do adequate preparation, it may be necessary to sandblast or water blast to get down to a sound surface. In most cases, just a thorough scraping and wire brushing will be enough.

Some paint removers are very effective if you have to strip off all the applied coatings. But paint remover is strong stuff. Follow the manufacturer's directions and recommendations.

Many surfaces require special treatment. Let's look at those you're most likely to encounter.

Glossy Surfaces

Glossy surfaces should be sanded, washed with a

solution of T.S.P., or treated with liquid sand paper. This provides a roughened surface or "tooth" for good adhesion.

Calcimine

If you have to cover a surface that's been painted with calcimine or a similar coating, start with a thorough washing. Adhesion to the calcimine isn't the problem. Paint will stick just fine on a calcimine surface. But applying paint over calcimine seems to release the calcimine from the surface *it* is on. That causes peeling.

Efflorescence

Efflorescence (calcium deposits) are common on masonry and concrete surfaces, especially surfaces below grade. Remove the efflorescence by washing with a 5% solution of muriatic acid. Allow the surface to dry thoroughly before applying paint.

Mildew

Mildew is a parasitic organism that's common in damp, poorly-ventilated areas that get little sunlight. It needs either high humidity or a damp surface to survive. But mildew can grow on many surfaces including wood, concrete, fabric, and some paints. If you notice a dirty, mottled discoloration on a wall or ceiling, it may be mildew. Recognize it by its blotchy, powdery appearance.

If you suspect mildew, examine the surface carefully. It may just be dirt rather than mildew. Dirt washes off easily. Mildew doesn't.

Any surface that has mildew must be completely sterilized before painting. Scrub the surface with a mixture of 2/3 cup T.S.P., 1/3 cup powdered detergent, and one quart of household bleach in a gallon of warm water. *Caution: wear rubber gloves and goggles when using this solution.*

Once you've removed the mildew, take steps to keep it from coming back. Add either a fungicide or mildewcide to the primer and paint you use when repainting the surface.

Chalking

Back in Chapter 5, I explained how chalking would affect your estimate. Now it's time to explain how to prepare a chalky surface.

Chalk is loosely-bound powder that forms on the surface of paint. Chalking happens when paint binder is destroyed by sun and moisture, when there's not enough binder to wet the pigment, or when the painter added too much thinner to the paint. Any of those conditions can make the binder disintegrate. The pigment is left exposed on the surface as a fine powder.

There are several degrees of chalking. Rub the surface with a finger or a dark cloth to judge how much chalk there is. Many paints can chalk mildly and still hold a good surface. Slightly chalky paint can resist moisture and weather for many years. Light chalking, especially on white paint, is desirable. It helps clean the surface every time it rains. Heavier chalking can cause a tinted paint to fade to a light pastel color.

Only fairly severe chalking is a problem when repainting. A heavily chalked surface has enough powder on the surface to make adhesion a problem. There's no firm surface for the new paint to bind with. This is especially true for latex finish coats which don't penetrate. They bind only to the surface powder, making the new paint no more durable than the chalk underneath. Severe chalking requires sandblasting or water blasting to remove the loose powder.

Moderate and light chalk needs only wire brushing or sanding to remove excess surface powder. But use a surface conditioner that will penetrate through the chalk and bind tightly to the surface below. Once you've done that, the new paint job will provide a firm base for later coats.

Wood Surfaces

Don't paint wood that's wet from high humidity, dew or rain. Let the moisture evaporate before you paint. Paint won't stick to a wet surface. Even some "green" lumber may have too much moisture to paint. Lumber with a moisture content of more than 15% is too wet to paint. Within 30 minutes of application, most paints won't be damaged by a light rain. But stop painting the exterior when rain starts.

Acoustical Surfaces

Acoustical surfaces are highly porous. They have thousands of tiny holes that absorb sound waves instead of reflecting them. Fill these pores and you destroy the sound-deadening quality.

If you have to paint an acoustical surface, start by cleaning with a vacuum cleaner. That removes most dust from the pores. If the surface has been painted previously and can withstand moisture without swelling or discoloring, wash it with soap and water and rinse with clear water.

Iron and Steel

When you're painting iron or steel, the most important thing is good surface contact. These metals rust when air and moisture get under the protective coating. That's why surface preparation is so important.

If you're painting very much steel, rent some blast cleaning equipment that will remove oxidation and loose paint. If only a small area is involved, use a medium or slow-drying primer with good wetting characteristics. Apply this after cleaning with a wire brush or power sander and rinsing with solvent.

If the existing paint is still in good condition, remove any loose paint and sand to feather broken edges around where the metal is exposed. Then spot prime before applying the finish coat.

Galvanized Metal

Galvanized metal doesn't rust as easily as iron and steel. But it's harder to paint because most paint doesn't adhere well to a galvanized surface.

Clean the galvanized surface of all oil, dirt, and grease. Then treat with a *vinyl wash pretreatment*. Remove all loose paint and rust spots on previously painted surfaces. When brushing or sanding, try not to wear through the galvanizing.

Aluminum

Clean with solvent to remove oil, dirt, and grease. Then treat with a vinyl wash pretreatment. If the surface is weathered, wire brush and sand to remove corrosion.

Stainless Steel, Chrome, and Nickel

Paint doesn't stick very well to these surfaces. Sanding with wet or dry paper will provide some "tooth" that makes adhesion more likely. Then treat with a vinyl wash pretreatment. Priming isn't needed.

Concrete Floors and Stairs

It's hard to put a durable coating on concrete floors and stairs. Usually the paint begins to crack and erode after a fairly short period of time — especially in heavy traffic areas. Concrete floors below grade are even harder to keep painted. If there's moisture coming up through the concrete from the underside, waterproofing is needed.

Here's how to test for moisture passage through a concrete floor. Place a piece of polyethylene sheeting (at least one yard square) on the floor and tape the edges down. Leave it there for at least 24 hours. If any moisture accumulates under the plastic, there's enough dampness passing through the surface to cause paint to peel. If so, waterproof the surface before painting.

Generally, concrete floors should be acid-etched before painting. Etching makes for better adhesion and neutralizes the alkalinity of the surfaces.

Preparation of old concrete floors may differ. If the floor already has a finish on it, and it's not peeling or lifting, just clean it thoroughly before painting. If the finish is cracked and peeling, scrape and remove all loose paint. You may have to sandblast to get a clean, even surface. On a badly damaged floor, acid-etching is necessary.

Always be sure that floors are thoroughly dry before you paint.

Primer Application

Many painters underestimate the importance of the primer. It isn't just a cheap first coat. The primer or undercoater is the foundation that supports the finish coat. Understanding that should help you understand the importance of primer. The best primer available for the job is going to be the best choice in every case.

Primer isn't intended as a finish coat. In fact, primer can be destroyed if it's exposed too long to sun and weather. If the finish coat is delayed until the prime coat becomes chalky or weathered, apply another coat of primer. Primer hardens as it ages, making it impossible for the finish coat to penetrate or adhere properly. The result will be peeling. Primer that has hardened can be restored with a light sanding.

Priming is always needed after patching plaster or wallboard. Spot prime the repaired area with a primer-sealer. When spot priming, feather the edges of the primed area and overlap the old finish enough to protect the adjacent area. If most of a wall was repaired, cover the entire wall with primer-sealer after spot priming.

Prep for Wallcovering

Start wallcovering application by washing the wall with a strong cleaner such as T.S.P. Then rinse the surface and seal with a flat, oil-base sealer. Cleaning and sealing are especially important on enameled surfaces in kitchens and bathrooms where the gloss finish probably has an accumulation of grease and dirt.

Mildew is always a problem with vinyl wallcovering because vinyl doesn't breathe. There's no way for moisture in the adhesive to escape except through the seams. This could take months. Most wall covering adhesives have a fungicide added by the manufacturer to prevent mildew. Wheat paste doesn't have an anti-mildew agent and is not recommended for vinyls.

Painting Vinyl Wallcovering
If part of your job is painting over vinyl wallcovering, begin by examining the vinyl. If it's still in good condition and is securely bonded to the wall, there's no reason to remove it. You can paint right over it. Here's how. Clean the wallcover as you would any previously-painted surface. Prime with an oil-base primer. I've found that it's a good idea to test a small area for adhesion. Wallcovering manufacturers use many different coatings. Most coatings accept paint, but some do not. Acrylic paints are usually a good choice, but vinyls may be required.

Painting Enamel over Wallcovering
If you have to apply enamel on a wall that's covered with vinyl wallcovering, it's usually best to remove the wallpaper first. Use a wallpaper steamer. Then remove all glue or paste residue from the wall. Sand and patch where necessary. You may also have to seal the wall.

You can apply enamel over some smooth, flat wallpapers. First seal the paper with an enamel undercoater. *Do not* use a latex or water-based undercoater. The water may re-wet the paste holding the wallpaper, causing it to lift.

Exterior Preparation
Up to now, we've been talking about preparation for interior painting. Let's go on now to exterior prep. We'll cover both stucco and wood siding and then pick up on a few special subjects like peeling, cracking and prep for metals.

Preparing Stucco for Repainting
If you're repainting an older stucco house, the first consideration is the type of paint that's currently on the stucco. Begin by testing for *water bonding cement paint (B.C.P.)*. It was common on stucco homes built in the 1930's and 1940's. B.C.P. is a primary cause of vinyl paint failure. As the B.C.P. deteriorates, it becomes very chalky and loses adhesiveness.

Even if B.C.P. is one or two coats below the exposed paint film, it may not be able to take the added weight of additional paint. Applying one or two new coats may cause all the paint to peel back to the old B.C.P. In a year or less your new paint job could deteriorate into a real mess. It wouldn't be the fault of the new paint. The problem started years before, when B.C.P. was used.

Here's what to do if you suspect B.C.P. First, rub your hand over the surface to feel for chalkiness. If the surface is chalky, make the following test. Drag a brass key across the surface. If it digs neatly into the paint but leaves no black mark, it's probably B.C.P. If the key leaves a black mark, the paint is either vinyl or an oil-base.

If the existing surface *is* B.C.P., sandblasting or thorough wire brushing will be the first step in repainting. If the surface is chalky oil or vinyl paint, remove the chalk with a wire brush and wash off all paint residue.

On an older building, the surface paint may be oil or vinyl, but there could be B.C.P. underneath. To check for B.C.P. below the surface, use a knife to scrape off all paint in a small area. Remove the paint right down to the bare stucco. Rub your fingers over the area to feel for chalkiness. If there's chalk against the stucco, the first coat was B.C.P. You'll have to sandblast before repainting.

Even if B.C.P. is chalk-free after being cleaned for painting, it's still not an acceptable painting surface. Eventually it will deteriorate, even after a new coat of paint is applied. The only way to be sure the new paint will last is to remove all the B.C.P.

Cracks or Breaks in Stucco
Repair all cracks and holes before painting. Fill gaps around windows, door casings, or where two materials meet, such as at the foundation line or where wood meets masonry. Use latex caulk to fill these holes. Check for cracks at the base of walls next to concrete drives and patios. Use latex caulk here also. A crack 1/16" or larger should be chiseled out to form a "V" shape so the caulk will fill the void and adhere firmly.

Preparing to Paint

Remove any loose or crumbling masonry. Then fill the hole with a powder or ready-mix latex-base stucco patching material. The latex types come in both smooth and textured finishes. If you use a dry powder, the mix will stick better if you apply a concrete or cement glue. Paint a little of this on the surface to be patched before adding the patching mixture. This may be the only way to make a good patch in a shallow hole.

Before using the patching material, dampen the hole with a sponge and clear water. This keeps the patch from drying too quickly and falling off.

If you use a non-textured stucco patching material, texture the area to make it match the existing stucco. Use a coarse fiber brush, like a scrub brush. Rub the brush over the partially set patch in a circular motion until the repair looks about like the rest of the wall. It isn't too hard to make a reasonably good match.

Remove any efflorescence on stucco or masonry with a stiff-bristled brush. Then neutralize the salt with a 5% solution of muriatic acid. Rinse thoroughly with clear water.

Preparing Exterior Wood Surfaces

Cracking and peeling paint are common on wood siding. In fact, it's expected after eight or ten years of exposure to the elements. Figure 7-9 shows cracked and peeling wood siding before my crew went to work.

You're not doing a professional job if any cracked or peeling paint is still in place when that first can of paint is opened. I recommend using a water blaster if there are more than 200 or 300 square feet of siding to prepare and the schedule allows several days of drying time. Water blasters are available at

Cracked and peeling paint on wood siding
Figure 7-9

Using the water blaster
Figure 7-10

most larger rental yards and are simple to use. Connect a garden hose to one end of the machine, turn on the water, switch on the machine and you're blasting.

Hold the tip of the gun at an angle to the surface being blasted. This helps water lift paint off the surface. Figure 7-10 shows the proper angle. You may have to spray some areas at several different angles to remove all paint.

The disadvantage to water blasting is that it leaves the wood saturated with water. You have to wait until the wood has dried completely before painting. A sweep with a high pressure blaster can inject water deep into old wood. It may be several days before this water finds its way back to the surface and evaporates. Paint too soon and you'll have a rapid repeat performance of the cracking and peeling you were called on to correct.

Usually the moisture will evaporate after two days of dry weather. If there's any doubt, many larger paint stores can provide a moisture meter to test for remaining water.

Your next step is to sand the wood surface. This creates a smooth, even finish and removes any old paint that was missed by the water blaster. Use a hand-held disc sander for this work. Sanding a large surface by hand takes too long and doesn't do as good a job as a disc sander.

On the first pass with the disc sander, use a rougher grade of sandpaper. Then finish up with a finer grade of paper. The grade to use depends on the condition of the surface. You may have to try a couple of grades until you find the right one. Note that rougher grade papers can gouge the wood. Be careful not to press too hard or dwell in one spot too long.

Preparing to Paint

Using the disc sander
Figure 7-11

Hold the sander with both hands and move it in the direction of the wood grain — not across the grain. Wear safety goggles and a dust mask when using an electric sander. See Figure 7-11.

Keep sanding until you have a smooth surface ready for finish paint. Then sweep off sanding dust and debris. The job's not done until the surface looks like Figure 7-12.

Fine-sanded wood surface
Figure 7-12

The next step is to spray or roll on the primer. But first be sure to cover all adjacent areas, including plants. See Figure 7-13.

Do good prep with the right tools, use good materials, and even the worst-looking siding will look like new. Compare Figure 7-14 and Figure 7-9.

New Wood Surfaces

Preparing a new wood surface is usually easier. But prep is still necessary. Start by removing all sap and wood splinters. Then sand in the direction of the wood grain. You may want to start with rough sandpaper. Final sanding should be with finer paper. Next, use a knot sealer to seal knotholes and areas where sap has accumulated.

All bare wood must be sealed with an undercoater. When the undercoater has dried, sand lightly and dust. Pay close attention to the rough edges of wood. That's where the wood is usually highly porous. Use enough undercoater on these edges so that they are still sealed after sanding.

Peeling Paint on Wood Siding

Excessive moisture can cause blistering, peeling, and discoloration. Figure 7-15 shows an example of this type of problem. If moisture is causing surface paint to peel, repainting (even after careful surface prep) won't solve the problem.

First you have to eliminate the source of the moisture. Water from rain, melted snow behind ice dams, or condensed water vapor may be getting into the siding behind the paint.

Do your client a favor. Find the source of the moisture and either correct it or suggest that your client correct it before more damage is done. A little searching should reveal where the water is coming from. Check for leaks in roofs and sidewalls. Are insulation, vapor barriers, and ventilation adequate? Make sure moisture from a clothes drier is vented to the outside. Check for leaky plumbing or deteriorated caulking around a bathtub or shower.

Repair the blistered surface by removing all loose paint. Apply a water-repellent preservative if you have wood joints that show damage. Allow the preservative to dry two days, or as directed on the container label. Prime bare surfaces and repaint, using blister-resistant paint.

Cross-Grain Cracking

After many repaintings with oil-base paint, built-

Rolling primer on exterior wood surfaces
Figure 7-13

up paint may get too stiff to stand the constant expansion and contraction of the wood below. The result will be cracked paint. The only remedy is to remove all paint down to the bare wood. Then prime the surface properly and repaint.

Intercoat Peeling
Intercoat peeling happens when the bond fails between two adjacent coats. It could happen if the primer and finish coat were incompatible or if there was a long delay between primer and finish coat. Intercoat peeling can also happen on any surface that's too smooth, hard, glossy, or oily. Latex paint separates from old paint that's even moderately chalky because latex doesn't penetrate very well.

There's no way to stop intercoat peeling once it's started. The only remedy is to remove the paint down to a sound surface. But it's fairly easy to prevent intercoat peeling. Be sure you apply the top coat within two weeks of applying the primer. Don't paint over glossy paint without roughing the surface with a strong detergent, steel wool, or fine sandpaper. Clean off oil or grease with mineral spirits or a household cleaner that contains ammonia. Remove surface chalk before painting with latex.

If you have any doubt that the surface is ready for painting, try this test. After prep is done, paint a small area with the paint you intend to use. Let the paint dry. Test for adhesion by trying to pull the paint off with Scotch tape. If it comes off, the surface isn't ready for painting yet.

Preparing to Paint

**Proper preparation makes a good job
Figure 7-14**

**Blistering and peeling paint
Figure 7-15**

Metal Surfaces

Avoid painting new galvanized metal surfaces until they've weathered for about six months. Galvanized metal comes from the factory with a residue of the manufacturing process and stain inhibitors that prevent good paint adhesion. Weathering tends to neutralize the surface, making it more ready to accept paint. If your client insists on painting galvanized metal right away, wash the surface with a mild acid such as vinegar. Rinse it thoroughly. Then apply a metallic zinc dust primer or other primer recommended for galvanized surfaces.

Now that you've completed the general preparations, it's time to pour the paint in a bucket and start the job. The next chapter will guide you through the process, from rolling flat paint on a wall to spraying a fine lacquer finish.

Chapter 8

Doing the Painting

The subject of this chapter is application — laying down paint by brush, roller and spray. We'll cover everything from rolling flat paint on a wall to spraying lacquer, with side trips through stripping and staining. There's lots of variety in the painting business. Every type of work requires different preparation, setup, materials, and application methods. This chapter hits all the bases.

Before I begin, let me dispel some illusions. Professional painters have long been slandered with the saying "Anyone can paint." I agree that anyone can paint, just like anyone can drive a nail, anyone can pull electrical wire, anyone can install plumbing fixtures and anyone can program computers. Anyone *can* load paint on a brush and wipe color on a wall. But only a real professional selects a compatible undercoat, maintains the correct coating thickness, knows how long to wait before applying the second coat, uses the right tools, and does the job quickly, neatly and profitably. Very few can do that. If you're one that can, you're a professional painter.

Painting, like any other trade, requires knowledge, practice, training, practice, experience, practice, skill, and more practice.

Anybody can go into the painting business. All it takes is a $1.95 brush and a calling card. This type of "painter" will often be bidding against you. His price will usually undercut your quote, making it more difficult to land good work with a reasonable profit. But the slapdash painting contractor isn't your competition. He lasts only until the community runs out of trusting fools. Past customers won't call him again. Every job he completes exposes his incompetence. Owners looking for a professional job won't hire him in the first place.

Think of painting as an art. Take pride in changing a rough, worn, old door into a spotless gem with the look and feel of glass. If doing that gives you a feeling of accomplishment, you'll have no trouble building a reputation as a competent professional painter.

Painting with Flat Paint

Let's start with the most basic job — applying flat paint.

Doing the Painting

I assume that you've estimated and planned the job, set up the job shop and done all preparation. You're ready for the next step, mixing the paint.

Mixing paint is easy if the paint is either fresh from the plant or has been agitated recently. If the paint's been sitting on a shelf for several weeks, have the paint store shake it up thoroughly. But always do a little last minute mixing to combine anything that has settled to the bottom. Stir it well and then pour it back and forth between two buckets. This is called *chasing* the paint. It's the fastest way to create a uniform consistency.

If the job requires two or more cans of the same color, mix the cans together in a large bucket. Paint may vary slightly in tone from can to can, even when it's the same brand and the same color. The only way to be sure you apply a single tone is to mix all cans together. This is called *boxing* the paint.

Thinning the Paint

Many paints have to be thinned for best application. Applied straight out of the can, the paint won't flow evenly. Experiment until you get a consistency that's just right for brushing, rolling or spraying.

If the paint's too thick, you'll find dot-sized spots of the previous color where the paint didn't flow out evenly to get coverage. But if the paint's too thin, there won't be enough pigment to cover. You'll have to apply an additional coat. Also, paint that's too thin produces a fine mist spray if you're using a roller. If the paint is water thin, this overspray will make a mess of the entire painting area. With a little experience, you'll know when paint consistency is right.

Note that some paints are intended to be thinned very little or not at all. The label on the can is your best guide.

Painting the Room

First, seal every surface that hasn't been painted before. Use a sealer recommended for the surface you're coating. Apply the sealer with a brush or roller, whichever you'll use for the final coat.

The color of the sealer should match the finish color. There are three ways to tint sealer. The paint store could add some tint to the sealer. You could mix some of the finish paint with the sealer. Or, you could tint it with some universal tints. Getting exactly the same color as the finish paint isn't important. Just get a color in the same neighborhood. That increases the covering capacity of the finish paint.

For crayon marks and other hard-to-cover stains, use the sealer your dealer recommends for its ability to cover stains. Don't be tempted to use the finish paint to seal difficult stains. Crayon and ink stains will still show through several coats of most wall paint.

When the sealer is dry, spot prime over areas that need special attention: patches in wallboard or plaster, crayon marks, water stains and the like. Use the same paint and the same brush or roller you'll use on the finish coat.

Brushing on Flat Wall Paint

This *is* very simple. Load the brush by dipping it into the paint one-third to one-half the length of the bristles. Pat the end of the bristles against the inside of the can to remove any excess paint that might drip. If you prefer, wipe the bristles against the inside rim of the can. But this usually wipes too much paint from the brush and decreases coverage before the next dip is needed.

The object here is to load as much paint on the brush as possible without creating a mess. Start with smaller amounts and work up to heavier loads once you have a feel for the brush, paint and surface being coated.

Your first stroke with the brush should be upwards. Then brush up and down as necessary to even out the paint. Wall paint dries flat (without a glossy reflection), so minor variations in film thickness won't be obvious. Painting with flat wall paint doesn't have to be perfect. Just avoid *painter's holidays* (areas where the paint did not cover) and heavy globs or streaks of paint (which form an irregular texture when the paint dries). Almost any other imperfection in stroking will be unnoticeable in the finished job.

Brush the paint out smooth. Begin each stroke in an unpainted area. Try to end each brush stroke in a painted area. Then lift your brush gently from the wall. If you just yank the brush away from the wall, you'll lift some of the paint off with it.

Use light brush pressure. Pressing down too hard makes the paint flow backwards towards your hand and smears the paint that's been applied. Gentle, even brush pressure lets the paint flow smoothly, with maximum coverage. But don't go overboard on this. Just experiment a little and see how much better a light touch works. This is especially true if you're using a good quality brush.

Tips on Cutting In

Roller jobs require brushwork, too. Corners and edges have to be painted by hand with a brush. That's called *cutting in*. Use a 2½" nylon brush to cut in where the ceiling meets the wall, at wall corners, around door jambs and windows, and around light fixtures and air conditioner vents. Here are some tips to make your cut-in fast and accurate.

Acoustic ceilings— You'll hear these called *popcorn* ceilings. They're very popular in modern tract homes. It's hard to paint the wall under a popcorn ceiling without getting paint on the acoustic material. This is especially true if the line where the acoustic ceiling meets the wall is irregular.

Even up the edge of a popcorn ceiling by running your spackle blade horizontally all the way around the perimeter. Scribing this line dislodges whatever acoustic material is touching the sidewall. It also creates a small gap that becomes the demarcation line between the ceiling and the wall.

Lightly load the brush with paint. Press the brush against the wall just below the ceiling line. Hold the brush at an angle so the forward edge of the bristles form a tip. Run this tip into the gap between the wall and ceiling. Move around the wall this way, laying down a band of paint around the perimeter of the ceiling. The band should be three or four inches wide so you can roll right up to the band without scraping against the ceiling.

Baseboards— Protect carpet with wide masking tape when you're painting baseboards. See Figure 8-1. Place the tape on the carpet so it extends up the baseboard about 1/4". Then tuck this overlap down into the joint between the edge of the carpet and the baseboard. Cover the tape and carpet with a drop cloth. This protects the carpet during painting.

When painting near the tape, keep your brush relatively dry. That way excess paint can't seep behind the tape. Figure 8-2 shows the technique.

Protect stained baseboard or baseboard that won't be painted with a strip of masking tape. See Figure 8-3. This masks off overspray from ceiling and walls. You'll need to cut in about three inches above the taped baseboard. See Figure 8-4. Keep the brush dry so excess paint doesn't leak under the tape and onto the baseboard.

Straight lines— On some projects you'll have a line on the wall or ceiling where one color ends and another begins. This is another detail that separates the pros from the amateurs. The line should be absolutely straight without overlaps or unevenness. That's not easy to do, as you must know if you've ever tried it.

I've never met a painter who could paint a good color joint freehand. I don't think you will either. Masking tape will help, but it's hard to get the tape straight. And on a porous surface, some paint will usually seep under the tape, ruining the line.

Say you're painting two parts of the ceiling different colors. Paint the first side of the ceiling without regard to the color joint. When that color's dry, measure carefully and draw a line exactly where the joint will be. With a thin flathead screwdriver or spackle blade, using a straightedge for accuracy, scribe a shallow groove into the ceiling along the line you just drew. When you paint the second side of the ceiling, just touch the edge of the groove with the tip of your brush. If you're careful, capillary attraction will pull paint into the groove but no further.

If the wall color and ceiling color are different, paint the ceiling first. Don't worry about overlap on the wall. When the ceiling is dry, scribe a shallow groove where the ceiling and walls meet. Then, with a dry brush, flow a light bead of paint into the groove. If you're careful, the bead will fill the groove but go no further. Paint the rest of the wall as usual.

Vertical straight lines— Here's how to handle a color joint on a wall. Hold a straightedge up vertically against the wall where you want the joint. Use a carpenter's level or plumb line to be sure the straightedge is vertical. Draw a straight line from ceiling to floor. Then use the same straightedge to scribe a shallow groove along this line with a sharp screwdriver or blade. Let a bead of paint flow into this groove.

Horizontal lines— Measure the height of the line from the floor. Mark the wall each three feet along the joint line. Connect these marks with a yardstick. If the floor isn't level (and they never are), use a carpenter's level to project a horizontal line on the wall.

Doing the Painting

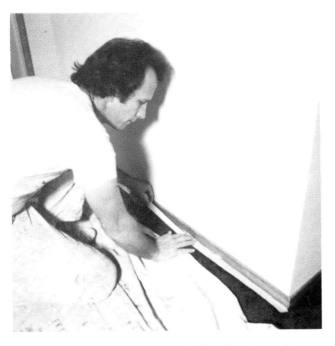
Protect carpet when painting baseboards
Figure 8-1

Paint baseboard with a "dry brush"
Figure 8-2

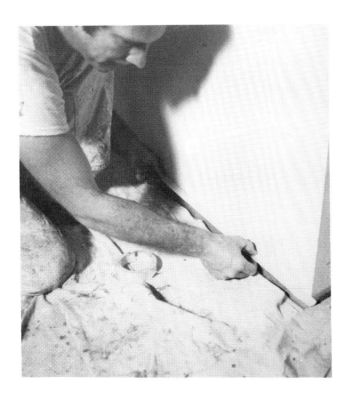
Protect baseboard with tape
Figure 8-3

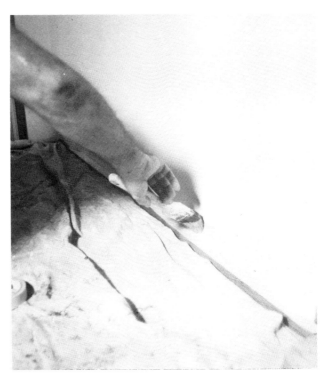
Cutting in above baseboard
Figure 8-4

Cutting in custom colors— When using standard colors straight from the manufacturer, the area painted by hand with a brush will usually match the area painted with a roller. You can cut in and roll out at will. The finished product will look about the same. This isn't always true with custom colors, especially deep tones such as a forest green. When using custom or deep colors, hand-painted areas probably won't match rolled areas when the paint is dry. The difference in tone is due to differences in texture left by a brush and roller.

So how do you cut in when applying custom color? You use two painters: one to cut in and the other to roll. One painter starts the cutting in and the other follows behind with the roller. It's important to roll the area while the cut-in paint is still wet. That tends to blend the two textures into one. So don't let the painter who's cutting in get too far ahead of the one rolling.

Take one more step when applying deep colors such as forest green. Again, use two painters working at the same time. But keep the cut-in as small as possible. Roll as close as possible to the ceiling, corners, and door jambs. When rolling up to the ceiling line, turn the roller sideways and get as close to the ceiling as you can. There will be a difference in tones here. But if you've rolled within an inch or so of the ceiling line, nobody is going to notice it.

A *long john* or *pencil* roller is good for getting close to the ceiling without getting paint on the ceiling. This is a 7" long roller with a diameter of only about one inch. It's usually attached to a long, thin metal roller handle. A long john roller is also excellent for painting hard-to-get-at areas such as cabinets and behind toilets.

Rolling Flat Paint

Every painter has used a roller pan to load a roller with paint. The pan works best if you use a roller grid in it. Fill the well of the roller pan half full of paint. Dip the roller until it's half covered. Run the roller up and down the grid two or three times. Dip the roller again and roll against the grid again until the roller is evenly saturated with paint. Don't get too much paint on the roller. That causes drips.

Roller pans have built-in disadvantages. Since they're flat and take up about a square foot of floor space, sooner or later someone's going to put his foot in one. They don't hold much paint. That makes frequent refills necessary. Every time you refill a pan, there's a chance of splattering paint on the floor. If you overfill, it's easy to slosh paint out when you move the pan.

I recommend that you use what the professionals use, a five-gallon bucket with a roller grid hanging from the inside rim. To load the roller with paint, just dip and roll on the grid as you would with a roller pan.

Buckets are a lot harder to step in than roller pans. You can move them without danger of spillage. And they hold lots of paint.

Rolling Paint in a Pattern

The most effective way to paint with a roller is to use a regular pattern. For walls, I recommend an "M" pattern. The first stroke is upwards, starting halfway between the floor and ceiling. Then roll down, up, and down again to complete the "M". To fill in the pattern, roll back across it the same way. Figure 8-5 shows the sequence.

The size of the "M" depends on paint coverage. Continue up and down strokes until you notice

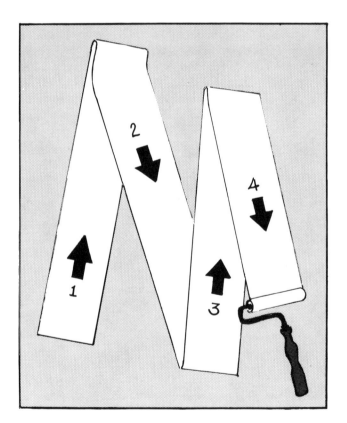

Rolling paint in an "M" pattern
Figure 8-5

that more effort is needed to move the roller. That's a sign that you're running out of paint. Now roll back across the "M", spreading the paint out evenly as you fill in the gaps.

Start in one corner of the room. Paint the top half of a small area of the wall. Then paint the bottom half. Continue across the wall, first painting the top of the wall, then the bottom.

Always use a light, even pressure when rolling. That keeps the thickness of the paint film consistent.

To smooth out excess paint buildup or lap marks (trails of paint left by the edge of the roller), roll the area once again just before loading the roller with more paint. Going back over the fresh paint with a relatively dry roller soaks up excess paint accumulation and spreads it to drier areas.

Lap marks can be a problem on a roller with a thick nap. But unless the lap marks are thick, they'll disappear as the paint dries. You can reduce lap marks by removing excess paint from the ends of the roller with a paint brush. The quickest way to get rid of excess paint on the roller ends is to apply the excess to the wall. Tilt the roller so only the edge is in contact with the wall. Then roll until the excess is gone.

Rolling paint on a ceiling is only a little different. I use the same procedure as on a wall except that my pattern is a "W" instead of an "M". Make the first stroke away from you. Some painters prefer an upside-down "V" or an "N". No matter what pattern you use, go back over the first strokes to even the distribution of paint. The idea in pattern rolling is to be systematic. First, get a lot of paint on the wall; then roll it out evenly. A word of advice: For ceiling work, use an extension handle on the roller and wear a hat.

Here are some general pointers when rolling paint:

1) The longer the nap, the less dipping needed and the faster the rolling. I wouldn't use anything

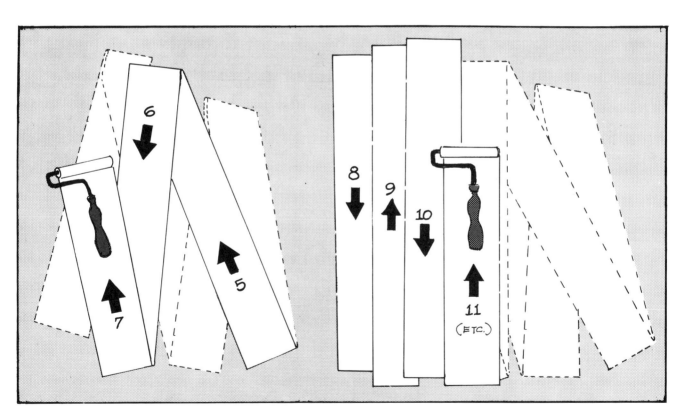

Rolling paint in an "M" pattern
Figure 8-5 (continued)

shorter than a 1/2" nap for smooth walls. Textured surfaces need an even longer nap. For top production, use a lambskin or synthetic lambskin roller.

2) Plan to spray paint acoustic ceilings. But if you have to roll, use a long nap (1½") and thin the paint. This is messy work, so be prepared.

3) When rolling flat latex paint in direct sunlight, cool the surface by spraying lightly with a hose before painting. This prevents the paint from drying too quickly, which ensures better bonding with the surface and allows more square foot coverage per gallon of paint.

Spraying Flat Paint

The key to spray painting is selecting the right equipment and using a good pattern. Many novice painters have trouble with spray equipment. Usually it's because they didn't follow the manufacturer's instructions. Clogging will be a constant problem if you don't read and follow the instructions that apply to the gun you're using.

It is absolutely essential to strain the paint and keep the paint bucket covered during spraying. Clean the machine carefully at the end of each day. *Never* leave paint in the machine, hoses or gun overnight. Paint that remains in the equipment very long will clog and corrode internal working parts. All good spray painting rigs are delicate pieces of equipment. Spare yourself frustration, downtime, and repair bills by cleaning spray paint equipment thoroughly and storing with the recommended mixture in it.

I use H.E.R.O. airless pumps. They require less maintenance and stand up to a lot of abuse.

Spray painters should always wear a spray mask and hood. See Figure 8-6.

Selecting a Spray Tip

It's impossible to do good quality painting at good production rates with anything other than the right spray tip. Here are my recommendations. For exterior latex on large unobstructed areas, use a size .021 tip. For interior and exterior latex flat and shake paint, use size .017 or .018. You can also use size .015 for interior latex and alkyd flat enamel. See Figure 8-7.

If the gun spits or splatters paint, your tip is either defective or partially clogged. Dismantle the

Wear a spray mask and hood
Figure 8-6

gun and clean the filter and tip. There are several ways to clean a spray painting tip. Clean the opening gently with a razor blade or push a thin wire or pin through it. Reverse the tip in the gun and blow out the clogging particles. For a more persistent obstruction, remove the tip and soak it in lacquer thinner. Then scrub the tip with a soft wire brush and push a thin wire through it. Be careful when doing this, as tips are quite fragile and can easily be ruined.

Spraying Patterns

Here's the right way to use a spray gun. Hold the gun perpendicular to the work surface and about 12" to 18" away from it. You want the spray pattern to be smooth and regular, so sweep the gun smoothly along the wall, starting the stroke *before*

Doing the Painting

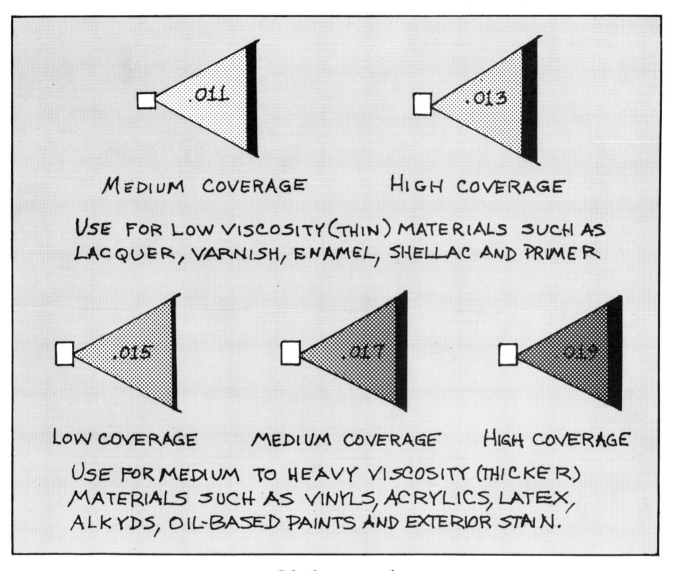

Selecting a spray tip
Figure 8-7

pulling the trigger. Release the trigger before ending the stroke. This prevents paint buildup at either end of the sweep. See Figure 8-8.

The second stroke should overlap the first stroke by one-half of its width. This ensures even coverage as you paint down the wall.

Test your spray pattern before starting on the wall. See Figure 8-9. If the pattern is spotty, you're probably not using enough pressure. Practice with your gun until you know just how much pressure to use.

Watch for drips and sags as you spray. If you find any, catch them with a roller or brush.

For acoustic ceilings, spray in a criss-cross pattern, front to back for the first coat and side to side for the second. Use very thin paint or special acoustic paint.

Painting with Enamel

In enamel work, the consistency of the paint is very important. This is especially true when painting doors, windows, and trim such as handrail, baseboard, or crown molding.

Enamel that has the right consistency goes on smooth. Paint that's too thick will leave brush marks and sags. That slows production. Paint that's too thin doesn't cover well, drips and runs.

You'll usually need to thin both oil-base enamel and latex enamel. Use the thinner recommended on the paint can label. The label will also tell you the maximum amount of thinner to use. I sometimes use a little more thinner than recommended, but only when I'm sure that more thinner is needed.

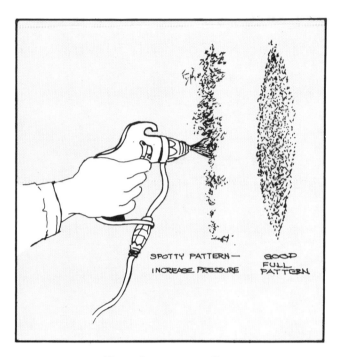

Test the pattern first
Figure 8-9

Correct spraying pattern
Figure 8-8

Here's how to test the consistency of enamel. Dip your brush in the paint, covering about one-half the length of the bristles. Remove the brush and hold it over the can. When enamel is the right consistency, it flows smoothly off the end of the brush. It shouldn't run off like water or sit on the end of the brush like glue. Figure 8-10 shows the right consistency for enamel. The final test of consistency is in the actual painting. You should be able to cover several inches (depending on the size of your brush) easily, and get good hiding capacity without sags or runs.

Always pour off a little of the paint before you begin thinning. That way you can thicken the mixture again if you get it too thin.

Some paint will thicken noticeably as the job progresses — especially when painting in a warm room or outside during a hot day. Heat evaporates the thinner in the paint. When necessary, add more thinner to restore the right consistency.

Oil-base enamel usually requires paint thinner, turpentine or a product called *Penetrol* for thinning. Penetrol is an oil-base additive that's especially useful when mixing color paints, because it doesn't reduce the effectiveness of the color pig-

**Thin your paint to the right consistency
Figure 8-10**

ment in the paint. It's also useful when painting in a hot room or on a hot day. The oil in the Penetrol won't evaporate like paint thinner, so it stays effective longer. It also helps provide a smooth, even painted surface.

For latex enamel, water is the most common thinning agent. But other thinners are available. Check with your paint supplier.

Painting Interior Trim

Many people will judge the quality of your work by the enamel on the trim, doors and jambs. Use extra care here, both during preparation and in painting. Even if you've done a good job on all prep work, take a few minutes to examine the trim closely. Any imperfections will show under the new coat of paint. So start with a surface as nearly perfect as possible.

Look for any loose paint. Scrape it away with your spackle blade. Pay close attention to all cracks and crevices. Check the area where the wall meets the door jamb. Watch for splinters. With a medium-grade sandpaper, sand any areas that aren't smooth. Run a bead of latex caulk into any cracks or splits in the wood. Then moisten your finger or use a wet rag to smooth out the bead. Make sure the surface is even.

Fill small nail holes with painter's putty. Apply the putty with your finger and then rub away the excess. Be sure the entire hole is filled. For larger holes or cracks, use pre-mixed spackle.

To fill a large hole that goes through to the interior of a hollow core door, use Fix-all and wire screen for reinforcement. The procedure is the same as described earlier for patching large holes in plaster or wallboard. Using Fix-all is easier, though. You can hold a screen patch in place for a minute or two until the Fix-all begins to harden. No pencil is needed. Sand carefully so the final coat is very smooth.

Apply a good quality enamel undercoater (sealer) to all patched areas and any raw wood. When it's dry, sand the entire surface with a fine sandpaper such as number 220-A. Dust carefully and wipe with a *tack rag*. This type of rag is available at most paint stores. It removes the fine sanding dust that a brush would miss. Let the dust settle before you start to paint.

If you're painting a new door, sand all splintered and rough edges carefully and then dust. Apply one or two coats of sealer. Sand lightly after each coat. That's important. Sealer can raise the wood grain, creating a rough surface. Be sure to seal the tops and bottoms of all new doors. After the sealer is thoroughly dry, sand with a fine paper and dust carefully.

Painting Doors

Let me emphasize the importance of cleanliness when painting with enamel. Even the smallest particles will show up when the paint is dry. That can ruin a good enamel job. There's nothing more discouraging than to discover several dirt particles in a freshly painted door. Be sure your brush, paint can, and paint are free of debris.

If the paint is new and fresh out of the can, it should be clean. But paint that's been opened and then left on a shelf for a while will probably have dried paint particles. Remove these impurities by straining the paint. Most paint stores sell strainer bags, but you can make one out of an old nylon stocking. Either will do a good job.

Place the strainer bag or nylon stocking inside an empty bucket. Make sure the bucket is clean. Stretch the top of the strainer over the lip of the bucket so it's secure. Pour the dirty paint through the strainer. Then lift the strainer, still holding it over the bucket. Let the paint run through until all that's left in the strainer is the debris. If the paint is thick, speed up the process by squeezing the strainer lightly with one hand.

If you're using a new brush or if the brush has been sitting unused for a while, clean it with the right solvent. If you're using an oil-base enamel, clean the brush in paint thinner. If the paint is latex enamel, use clean water. This cleaning will remove dust, dirt and any loose bristles. Spin the brush to dry the bristles.

Now, you're almost ready to paint. But first, place a drop cloth completely under the door. Secure the door with a stop so it won't move while you're working. Finally, pour about a quart of paint into a clean one-gallon can.

When painting any type of door, start with the edges. If the door is a different color on each side, follow this rule to determine the color of each edge: *The edge that's visible from a room should be the same color as that room.* See Figure 8-11.

Doors with flat surfaces: Now we're ready to get started. Follow along on Figure 8-12 as we go through the instructions.

Paint the back edge of the door first (Step One in Figure 8-12). Dip your brush about one inch into the paint. Pat the brush on the inside wall of the bucket until paint no longer drips from the bristles. Hold the brush so the bristles are facing the door edge. Start at the top and paint a straight line to the bottom. After you've painted the entire edge, go back and wipe the hinges with a cloth wrapped around a spackle blade. Then use a rag to clean any excess paint from the side of the door you're *not* painting.

Now load your brush with paint. Start in the upper left corner and apply paint about two feet down and halfway across the door. See Step Two in Figure 8-12. Then, using the flat side of your brush with the edge facing down, and starting at

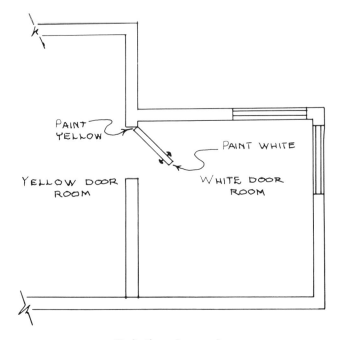

**Painting door edges
Figure 8-11**

the bottom of the painted section, brush up toward the top of the door (Step Three in Figure 8-12). This is called *laying off.* It smooths out the painted surface. Laying off is the secret to getting a good enamel job. It leaves enamel with a smooth glass-like finish that shows you're an expert professional painter.

When you've laid off this first section, proceed to the unfinished portion at the top right of the door. See Step Four in Figure 8-12. Once the paint is applied, lay off this section. Continue with this procedure until the entire top half is completed. Then lay off the whole top half in several even strokes, starting from the left and moving to the right. See Step Five in Figure 8-12.

Now repeat the procedure until the bottom half is complete.

When laying off the bottom half, you will be brushing directly into the freshly painted top half. It's very important here that you continue the upward stroke a few inches into the top section. This makes the top and bottom sections flow together. See Step Six in Figure 8-12.

Try painting an old board or piece of plywood using the procedure I've just described. With a little practice, you'll be surprised at how good the paint job looks.

Doing the Painting

Step one: Paint back edge of door.

Step two: Apply paint about two feet down and halfway across.

Step three: Lay off painted section.

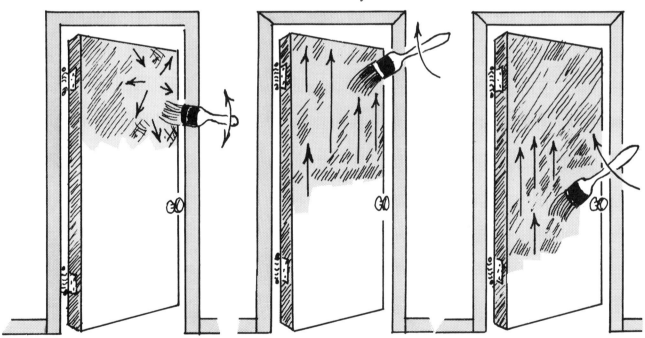

Step four: Repeat steps two and three on right side.

Step five: Lay off the whole top half.

Step six: When laying off the bottom half, continue the upward stroke a few inches into top section.

**Painting flat-surfaced doors
Figure 8-12**

Don't let anything interrupt work when you're enameling a door. Timing is important. Move from one section to the next so that none of the painted sections dry before you complete the laying off. And don't spend too much time admiring your work. Once you start a door, finish it promptly.

You will see here how important it is to have the right consistency in the paint. Only if the consistency is right can you keep a wet edge on all painted sections without the paint sticking or running.

Occasionally the job may become a mess. Either some dirt got into the paint or the consistency wasn't right or an unavoidable delay made you stop in mid-stream. When that happens, there are two choices: You can either remove the paint by wiping it with a rag soaked in the proper solvent, or let it dry and start over again later.

It's faster to roll the enamel on a flat door. But rolling leaves an uneven *orange peel* texture. Try combining roller speed and brush quality! Roll the enamel on and then lay it off with a brush.

Carved, panel and molding doors: The procedure for painting carved doors is very similar. Follow along on Figure 8-13. Start with the edges. Then paint the moldings or carvings, using smooth, short strokes. You may have to poke the end of your brush into fine details in carvings or moldings to get complete paint coverage.

Once you've painted all of these areas, begin on the insert in the middle of the door. Paint in long, smooth strokes from the middle *out to the sides,* top and bottom.

Next, go back to the top of the door. Using long smooth strokes, paint the flat surfaces at the top, sides and bottom. Follow this procedure until you've painted all the surfaces between insets or moldings. See Step Four in Figure 8-13. Check constantly for drips. The trick here is to paint fast enough so that the paint keeps a wet edge. It's hard to get a perfect match when painting into an area that has already dried. Work at a good consistent rate so it all flows together.

Here are two points to remember. First, when painting carved, molding or inset areas, get as little paint as possible on the flat portion of the door. If you lap into flat areas and if the paint dries too much, you'll have *brush drag* when you paint the flat part. Second, watch for drips in the molding and inset areas for several minutes after painting is completed. The more intricate the area, the more likely a drip.

When you see a drip, dry your brush against the inside lip of the paint bucket. Then use light, short strokes to brush the drips away.

Tips for Painting Doors

Here are some points to remember when painting doors.

• The easiest way to paint a door is with the hardware off. But if you don't remove the lockset, mask it with tape and then paint a smooth line around the knob. Then brush *away* from the knob, using a light touch. This will blend the paint around the knob with the rest of the door.

• Note directions on the paint can regarding temperatures. Don't paint if it's too hot or cold. Temperature has a major effect on the finished job. If it's too hot, the paint sets up before you can brush it out smoothly. If it's too cold, the paint won't set up at all. You'll find drips or sags occurring as long as 30 minutes after you've painted the door.

• Use a clean, soft-bristle brush. Be sure the bristles are even. A 2½"- to 3"-wide brush is about right for doors.

• When painting near the bottom of a door, be careful not to pick up debris from the floor or drop cloth.

• Don't paint while dust is floating in the room.

• Always paint the front door of a home before noon. With a little luck, it will dry before the family comes home for the day.

• When finished, wedge something under the door so it doesn't close until it's completely dry.

Painting Enamel Door Jambs

When painting a door jamb over carpet, apply 2"-wide masking tape to the carpet where the jamb and the carpet meet. Mash the tape down so you can paint right to the bottom of the jamb. No unpainted jamb should be visible when you remove the tape. The masking tape protects the carpet as you paint the bottom of the jamb, and helps prevent debris from the carpet getting on your brush. Let the paint dry overnight before removing the tape.

Doing the Painting

Step one: Paint door edges.

Step two: Paint edges of molding or carved areas.

Step three: Paint remainder of molding or carved area.

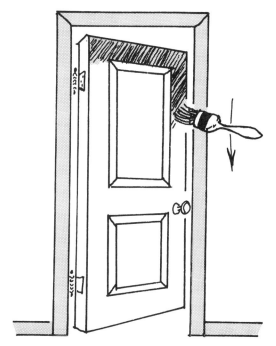

Step four: Paint all remaining areas between or around insets or molding.

Painting doors with molding, carving or inserts
Figure 8-13

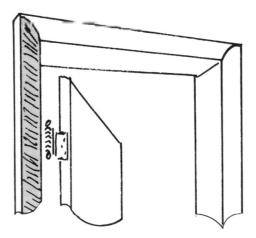

Step one: Paint left side, inside and outside.

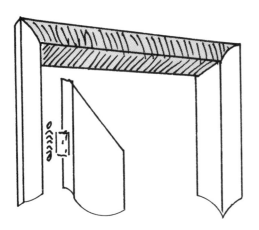

Step two: Paint top, inside and outside.

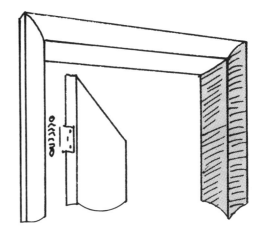

Step three: Paint right side, inside and outside.

Painting door jambs
Figure 8-14

Here's the right painting sequence for a jamb. Start painting at the left vertical section. Paint the inside portion of the jamb first. Be sure the paint flows into any grooves or crevices. Then paint the section that faces out, toward you. See Step One in Figure 8-14. Use the usual enamel technique on a door jamb. Get the paint up there first and then lay it off with even strokes.

Next, paint the horizontal section of the jamb, following the same procedure. Paint the inside first, then the outside. See Step Two in Figure 8-14. Finally, move to the right side of the jamb and do it again. This way you're always painting into the wet edge.

There are two ways to paint the outside edge of the door jamb, where the jamb meets the wall. You may hear these called the *French method* and the *Hollywood method*.

In the French method, the outer edge of the door jamb is painted with enamel. See Figure 8-15 A. To paint this edge, use only a small amount of paint on the brush. That keeps excess paint off the wall. Hold your brush with the edge toward the jamb edge. Paint from top to bottom. Paint this outer edge before painting the rest of the jamb.

Using the Hollywood method, the edge meeting the wall gets painted with the same paint as the wall. See Figure 8-15 B.

Doing the Painting

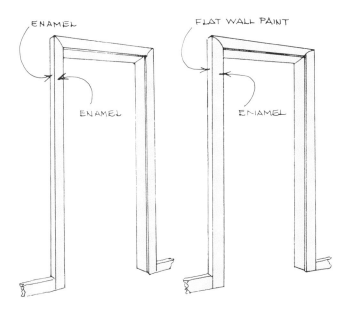

A) French method   B) Hollywood method

Painting the edge of the door jamb
Figure 8-15

The French method is preferred. It adds a nice touch to the finished product. But it also takes longer and it's harder to do a first-rate job. Figure 8-16 shows a jamb with the edges painted by the French method.

Painting Window Interiors

Windows in older homes are often stuck shut. Either they were painted shut or the weather has caused them to warp and bind. A stuck window isn't your responsibility. Don't even consider working on a stuck window without getting your client's approval. Some windows won't come free. Forcing the issue could mean replacing a window at your own expense — even though the old window was worthless. Just tell your client that the window is stuck shut. Your client will then do one of four things:

1) He'll open the window himself (or wreck it in the attempt).
2) He'll tell you to paint the window in the closed position.
3) He'll hire a carpenter to replace the window.
4) He'll ask you to try to open it (and absolve you of responsibility if it is damaged in the attempt).

No matter which alternative your client selects, you won't end up hanging a window or setting glass at your own expense.

Here's how to free a stuck window — if that's your client's choice. Insert a clean spackle blade or window-opening tool at the point where the sash is binding against the frame. This will break the paint seal. Then gently tap the window with the heel of your hand. You may have to insert the spackle blade at several points until the window finally loosens up.

Windows deteriorate faster than other interior woodwork because they're exposed to light,

Door jamb with edges painted by the French method
Figure 8-16

moisture and weather. That's why windows need more prep work. As with any enamel work, take a minute or two extra to be sure the window prep is done right.

Remove all cracked and loose paint. Fill small holes and cracks with painter's putty. Spackle larger holes. Smooth the patch and sand carefully. Then use a good primer to paint *all* patches and *all* unpainted wood. Allow plenty of drying time. It isn't necessary to prime patches made with painter's putty. Give the entire window a final sanding with fine grade paper (220-A). Always sand *with* the grain of the wood. Then do a good final dusting. If the window glass has dried paint from a previous painting, scrape this away with a single-edge razor blade.

Sand new windows to remove all splinters and roughness. Then dust carefully and apply a good enamel undercoat. Be sure to seal tops and bottoms of the windows. This will keep moisture out of the wood and prevent warping. It's a good idea to use two coats of undercoat on a new wood window.

Sash windows: Painting sash windows is slow work. Muntins divide the glass area into small panes. Each pane of glass is called a lite. When painting this type of window, start on an inside edge near the top of the window. Work from the top down. Use a 1½"- to 2"-wide brush. Sash brushes, designed especially for windows, are angled at the end.

Look at Figure 8-17. Step One is painting all horizontal and vertical muntins. Dip just the tip of the brush to keep over-painting to a minimum. But be sure all corners are completely covered. Next, paint the sash around the perimeter of the glass. That's Step Two in Figure 8-17. The last step is to paint the window frame and sill.

The job will go faster if you clean paint from the window as you work. After painting two or three lites, and while the paint is still wet, scrape all paint off the glass with a single-edge razor blade. Keep a rag in one hand to wipe the blade. This will save many hours of scraping later.

Double hung windows: Every surface on a double hung window has to be painted — even surfaces that aren't exposed to view when the window is closed. Figure 8-18 shows how to do it step-by-step. Start by pushing the bottom sash up and the top sash down. Paint the bottom two inches of the top sash. Then return the window to the closed position. Complete painting of the top sash. Finally, paint the lower sash, frame, stool and apron.

Don't paint the underside of the upper window sash with interior paint. It's exposed to weather and moisture and requires exterior paint. And since it faces the exterior, it should have the exterior color on it.

After the paint has dried completely, open and close the window several times. This will eliminate any sticking.

Exterior Doors, Jambs and Windows
Use the same procedure for painting exterior trim as for painting interior trim. There are two differences, however: preparation and materials.

Once again, the better the preparation, the better the job. On many exterior jobs, preparation of the wood trim is the major part of the job. This is especially true of wood exposed to late afternoon sun. It's easy to spend several hours preparing a single badly-weathered door. Take the time to remove all peeling paint and repair split and cracked wood.

Always use paint and patching material intended for outdoor use. Paint exposed to weather deteriorates much faster than paint used indoors, even with proper preparation, good painting technique and the right materials. With poor preparation, careless painting and the wrong materials, the paint won't make it through next spring.

Here are some tips for painting exterior wood:

• Avoid painting in direct sunlight, especially on hot days. The paint dries too fast and gets *ropey*. Shade the surface with a tarp, if necessary. If you have to apply water-based paint in the hot sun, cool the surface first with a fine water mist. And then work fast.

• You may have to thin your paint a little more than usual on hot days in order to get a good flow at the brush overlap. Penetrol will help oil-based enamel to flow better on hot days.

• Don't leave your brush out of the paint or its solvent for very long. It can harden in the sun very quickly.

Doing the Painting

Step one: Paint all muntins.

Step two: Paint the sash surrounding the panes.

Step three: Paint the frame, trim and sill in the order of sequence shown.

Painting interior windows
Figure 8-17

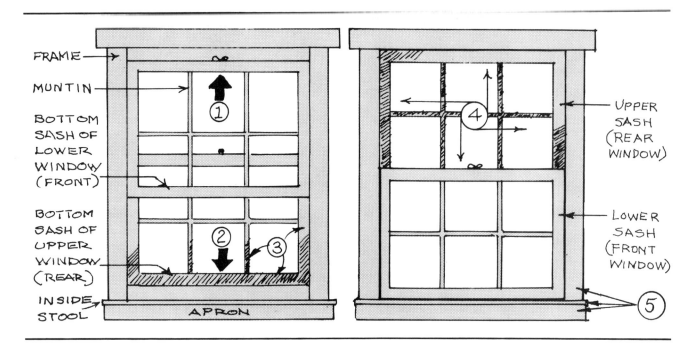

Step one: Push bottom window up.

Step two: Pull top window down.

Step three: Paint bottom of lowered window sash and muntins (at least 2").

Step four: Return windows to closed position and complete painting of windows, working from middle outwards.

Step five: Paint frames, inside stool and apron last.

**Painting double hung windows
Figure 8-18**

- When it's hot, don't leave any brush sitting in a bucket of paint for very long. If you can't find a cool, shady location, soak a heavy rag with the proper solvent. Lay this rag over the can with the brush handle sticking out. This is only a temporary solution. Don't let it sit for much longer than 30 minutes. If you're taking a longer break, put a lid on the paint and soak your brush in solvent.

- On exterior windows, always check the condition of the sash glazing (if glazing was used to hold glass in the sash). Older windows usually have at least some cracked or pitted glazing. Scrape out and remove deteriorated glazing putty. Replace it with new glazing, then paint.

- Check badly damaged wood for termites and dry rot. If it's easy to poke a screwdriver through a window or door frame, deterioration has reached the point where replacement is needed. This isn't your job. But it's good professional practice to inform the owner when you find a serious problem he may not be aware of. It also protects you. Repainting rotting or termite-infested wood is an exercise in futility. And when, very soon after, the paint is lying on the ground or the rot is coming through, *you're* going to be the one blamed and bad-mouthed. You don't need this if you're trying to build a reputation as a quality painter.

Painting Cabinets

Preparing cabinets for painting is like preparing any wood surface. The major difference is that kitchen cabinets will usually have a film of grease. Wash down greasy cabinets with a solution of T.S.P. and warm water. You may have to wipe with paint thinner on heavily soiled areas. Remove all the cabinet pulls. Take out the drawers and

Doing the Painting

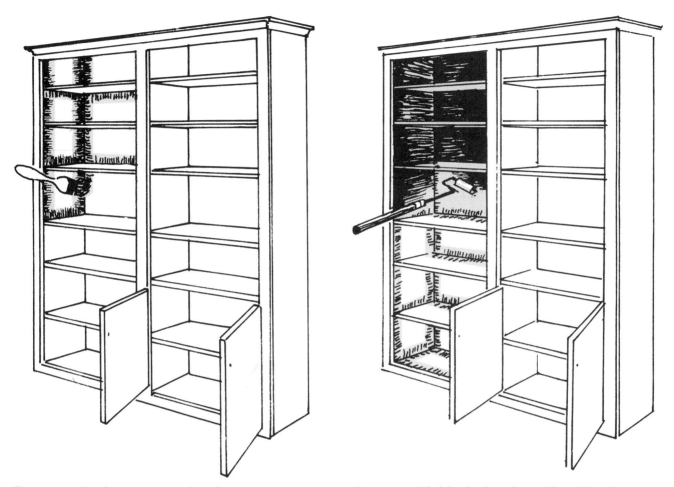

Step one: Cut in corners and ends.

Step two: Finish the interior with a 7" roller.

Painting cabinets
Figure 8-19

stand them on end.

If you're going to paint the inside of cabinets, start there first. Dust the shelves, starting at the top and working down. Be sure to get dust out of the corners. Painting the inside of cabinets with a brush is awkward. Consider using a small roller instead. My choice is a 7" roller handle and roller cover. You might prefer a long-handled "long john" roller. With either roller, use a two-gallon bucket and 7" roller screen for easy dipping.

Figure 8-19 shows how to paint cabinet interiors. Step One is to cut in with a brush at the corners and at shelf-to-wall intersections. A 2" brush works best. Now get a medium load of paint on your roller. Roll out the excess on the roller screen in your bucket. Start painting at the top of the cabinets and work your way down. That's Step Two.

When you're finished rolling the interior, look over the cabinet for holidays and drips. Do any needed touch-up while the paint's still wet.

Next comes the cabinet exterior. Start with the frame (Step Three in Figure 8-19). If the hinges are to remain unpainted, paint around them very carefully. You may need a 1" brush to do this.

Finally, paint the cabinet doors. Start with the door backs, then do the door fronts. Occasionally check back over the areas you have painted for drips or runs.

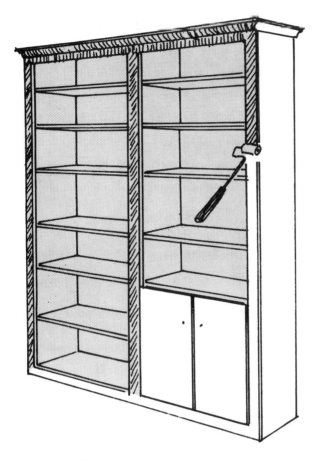

Step three: Paint the cabinet frame.

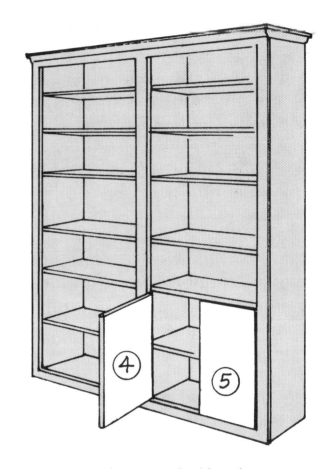

Step four: Paint the backs of cabinet doors.

Step five: Paint the fronts of cabinet doors.

Painting cabinets
Figure 8-19 (continued)

Staining

There are three types of staining. You may do one, two or all three types. All require a good knowledge of stains and wood. The major differences are in the skill of the craftsman, the quality of materials, and the time needed to do the job.

Fine Custom Staining and Refinishing

This is the masterwork of the painting trade. It takes a lot of study and practice to do custom staining. The client who requires this type of work is paying top dollar for a knowledgeable professional who will take the time to do it right. Quality is the only consideration. There's no compromise of standards. Entire books are written on this subject alone. Be sure you can really do this type of work before promoting yourself as a custom stain man.

Homeowner Staining

This still requires a good working knowledge of stains and wood. It's usually done for a client who wants a good professional job in his home but isn't willing to pay top dollar for exquisite work.

New Construction Staining

This is usually done in new apartments, condos, and office buildings. Like all staining, it requires knowledge of stains and wood types. But emphasis is on volume production, not exceptional craftsmanship. This doesn't mean that work is haphazard. There's just more of a balance between quality and speed.

The type of work you do depends on what the client wants and is willing to pay for. Find out exactly what your client expects. If a general contractor hires you to stain wood in 12 new apartments, he may be willing to pay for only one coat of stain and one coat of varnish or lacquer. Obviously, this isn't custom or homeowner grade staining.

This chapter has all the basic information you need to do good stain work. But I can't provide the practice a painter needs to become proficient in staining. That's up to you.

Equipment for Staining
Here are the special tools needed to handle staining:

Sandpaper: You'll need several sheets of medium grade and fine grade sandpaper. Use the medium grade for rough edges, splinters and removing old stain. Use the fine grade for touch-up and final sanding. Figure 8-20 is a guide to selecting the right sandpaper for stain work.

Rags: Be sure they are lint-free and clean.

Vaseline: Coat your hands before beginning work. This protects the skin and makes washup quicker.

Tack cloth: This removes the fine dust particles before varnishing.

Wood putty or painter's putty: Although there are several kinds of wood putty on the market, regular painter's putty is easier and faster to work with. A lot of wood putties don't take stain as well as the wood itself, and seal the surface adjacent to the hole patched. The wood around the patch will be a lighter color than the rest of the wood when staining is finished.

Plain painter's putty works very well. Tint it to match the stained wood as closely as possible. Standard universal color tints work well for this. If the putty gets too soft or wet from the tint, mix a little corn starch into the putty. This removes excess moisture.

Getting Started
First, prepare the working space. Cover the floor, tile and fixtures. Remove all the knobs and handles

Use	Grit number	Description
Rough sanding	80	Medium
Preparatory sanding (hardwood)	120	Fine
Preparatory sanding (softwood)	100 or 120	Fine
Finish sanding (hardwood)	180 to 220	Very fine Extra fine
Finish sanding (softwood)	220	Extra fine

Sandpaper chart
Figure 8-20

from the drawers and doors. Take out all drawers and removable shelves. Set all nails below the surface of the wood with a nail set and a hammer. If you see excess glue on the surface, remove it by sanding with medium paper. Glue residue will repel the stain.

Next, sand all rough surfaces, splintered edges and marred or stained areas. A 120-grade sandpaper will remove most hard-to-remove stains, and 220-grade will handle just about everything else. Use smooth strokes, following the grain of the wood. Hand-held "palm sanders" can be used on cabinets, but they may leave deep scratches if used with too much force or too rough a grade of sandpaper. So practice on some scrap wood if you've never used one before.

Finally, dust the cabinets with a dusting brush or blow them clean with an air compressor. Now you're ready to begin.

Sealing the Wood
Some types of wood have to be sealed in order to absorb stain evenly. Softwoods have a tendency to soak up more stain in some areas than others. This makes the final color blotchy. Softwood such as pine, birch and fir should always be sealed first.

When sealing wood, be careful about using sealer straight from the can. The surface may be sealed off too thoroughly and won't absorb enough stain. Then the color may be too light. I recommend using three to five parts thinner to one part sealer on soft woods.

Hardwoods such as oak, maple and walnut usually don't need sealing. They produce an even, rich color. But if you're worried about the quality of the surface, go ahead and seal first. If you seal hardwood, use eight to ten parts thinner to one part of sealer.

If you have any doubts about how much to thin your sealer, do some tests on scrap wood to determine exactly what consistency of sealer works best for that type of wood. *Don't experiment on the wood you're staining.* Once the sealer dries, it's extremely difficult to remove. If you're lucky, sanding may get it off. If not, you may have to strip the wood. Otherwise, you might have to apply two or three coats of stain to get the desired color, and that could obscure the grain pattern of the wood.

Apply your sealer with a clean brush and allow it to dry completely before staining.

Staining New Wood
New wood means wood that has never had any type of finish — stain, varnish or paint. Before staining new wood, test a small sample. Use a scrap of similar wood or find an area of the item to be stained that won't be visible when the job is complete. Apply a small amount of stain as a sample. Have your client approve that sample before proceeding.

Some wood has blemishes that don't appear until stain is applied. But you can expose blemished areas by spraying with alcohol. Alcohol reveals any spots and then evaporates quickly without affecting the wood. This gives you a chance to sand away blemish spots before staining begins. (This trick is particularly valuable when you're working with wood that has been stripped.)

Picking the right stain will make the job easier. Lacquer-based stains dry faster, giving you less margin for error. They're also harder to touch up. Oil-base stains give more depth, are richer in color and produce better highlights. Because they dry slower, they're easier to work with.

Penetrating wood stains are absorbed deeply into the wood and are generally used on interior surfaces. Pigmented stains contain color pigments which don't penetrate as deeply. Using pigmented stain can reduce color variations in softwoods such as pine and red fir, since it doesn't penetrate as much. Do some experimenting with the stain you select before proceeding with the job.

My experience is that transparent stain can be thinned much more than the manufacturer recommends. Just be careful not to get the color too light. You can also tint the stain with universal tints to get exactly the color you want. But be sure to tint a batch big enough to do the whole job, with a little left over for touch-up.

With patience and experience, you can learn to match existing finishes using universal tints in transparent stain. To get an aged, yellow look, add raw sienna to the finish coat. Another trick is to use two stains. Add tint or stain to the sealer coat to match the background color of the existing stain. Then stain over it with the color you want.

Using a brush or cloth, apply the stain evenly. But don't stain too far ahead before starting to wipe. Let the stain set for a few minutes. How long you'll let it set depends on the heat, humidity, and how much color you want.

For rough or sandblasted wood, you can spray stain, using a size .015 or smaller tip. If you thin the stain and flood the surface, about 90% of the stain will be absorbed. If you've sprayed correctly, there will be very little excess stain to wipe off. Proceed with caution the first time you try to spray stain, until you're sure your specific method is working.

If a blemish appears, sand it out with number 220 sandpaper while the stain is still wet. Sometimes you'll get a *fish eye* where the stain beads around an oily spot on the surface. When that happens, wipe the stain off and clean the surface with lacquer thinner and sandpaper.

The next step in staining is the wiping. While the stain is still wet, wipe with a clean cloth. Be careful not to remove too much stain around the edges. Wipe stain away until the natural grain of the wood comes through and you get the color expected. Many different effects can be produced in this wiping stage. When wiping is finished, let the stain dry completely.

Your next step is to apply sealer over the stain. If you're using a lacquer finish, apply a lacquer sanding sealer. If you choose varnish or polyurethane for the finish, use high-gloss for the sealer. This step seals the stain in the wood, gives you a sealed surface to sand without marring the stain, and prepares the wood for the finish coats.

After sealing, sand well, dust, and apply the finish coats.

Staining Over an Existing Finish
This applies to wood that has been previously stained and finished. The first step in staining old wood is to clean it thoroughly. Use T.S.P., rinse it off with water, then let the surface dry completely.

Next, search for places where the old finish has worn through. These spots have to be blended in before staining can begin. Clean each spot thoroughly and sand lightly with fine paper. Tint the sealer with a little bit of stain that matches the old finish. Apply this to worn or bare spots. Let the sealer dry and then sand lightly. This restores the finish so staining and sealing can begin again. If you stain before touching up the bare spots, the exposed wood will absorb too much stain and turn out darker than other areas.

When this is complete, sand the entire surface with a fine grade paper. Then dust, and use the tack cloth.

Apply new stain directly over the old stain and finish. Let it set for a few minutes. Using a clean brush or rag, gently brush or wipe the stain until you get the effect you're looking for.

You can do some experimenting when staining over an old finish. The stain won't soak into the surface as it does with unfinished wood. If you don't like the effect or color, start over. Simply wipe the stain off with a rag dipped in paint thinner.

Let the finished stain dry overnight before applying a clear finish.

Applying Varnish or Polyurethane
There's a lot of controversy over which is best, varnish or urethane. I give the edge to urethanes because of their durability and flexibility. Survey the shelves of your paint store. You'll find that the urethanes far outnumber the varnishes. Whichever you choose, the technique is the same.

Sand the surface with fine paper. Fill all nail holes and cracks with tinted putty. Next, dust with a brush, let the dust settle, then wipe the surface with a tack cloth.

Satin and semigloss finishes should be stirred, *never* shaken. Shaking produces bubbles which may show up in the finish later on. Gloss finishes should never be stirred *or* shaken.

Using a clean brush, apply thin, even coats. Thick coats have a tendency to drip or sag and leave brush marks as the finish dries. Use smooth strokes, keeping a wet edge on all areas. Make your final strokes go from end to end for an even finish.

You can also spray varnish or urethane, using a .011 or .009 tip to apply very light coats.

For good protection and a richer look, use at least two coats. Three coats are even better. Leave plenty of drying time between coats. Sand lightly with a *very fine* paper between coats for the richest look.

I recommend always using high-gloss urethane for the first coat, no matter what the final finish will be. High-gloss seals completely and holds finish coats away from the surface, giving the highest sheen. For that extra touch, hand rub the final coat with grade 0000 steel wool and paste wax. Apply the wax sparingly and let it dry before buffing with a soft, lint-free cloth.

Applying Sprayed Lacquer
Stain the surface just as you would for a varnish or polyurethane finish. But there's an extra step here. For safety, turn off all pilot lights in the area and open windows and doors to get as much ventilation as possible. Lacquer vapors are very volatile and can explode if allowed to build up in a closed room. Not only pilot lights, but the spark from a light switch or cigarette lighter can cause lacquer vapors to explode.

For spraying lacquer, use the finest tip possible, probably a .011 or .009. Apply several light, even coats, using a sweeping motion. Never hold the spray gun in one position. That's sure to produce runs in the surface. Avoid spraying lacquer when the weather is damp or when it's raining. It will cloud up on the surface being sprayed. If it does cloud up, you'll have to respray when the weather is dry.

Flat and satin finish lacquer can be applied straight from the can with an airless gun. But for semigloss or high gloss, thin the lacquer as follows: 50% lacquer, 40% lacquer thinner and 10% lacquer retarder. Otherwise you'll get an uneven orange peel texture to your finish. A conventional spray gun also works well for gloss finishes.

Don't spray lacquer on exterior surfaces. It will crack, chip and peel in just a few months.

Although spraying is by far the best, brushing lacquer is available if you can't spray. It has two major disadvantages. First, it sets up so quickly that there's very little margin for error. Second, you must wear a respirator mask while brushing to protect against dangerous fumes. Stay away from brushing lacquer unless it's absolutely necessary.

Stripping

Stripping is a very messy, dirty and unpleasant part of the painting business. Avoid it whenever possible. Consider taking furniture, doors and cabinet doors to a professional stripping service. They dip the wood in huge vats of stripper and then scrub it clean. But there will be times when some stripping is unavoidable.

With stripping, you never *really* know what to expect. I recommend that you bid these jobs on a time-and-material basis. If you can't do that, bid high enough to cover the risk and mess involved.

You may have to strip a surface several times, depending on how many coats of paint, stain or varnish have to be removed. Even with several strippings, it's sometimes impossible to remove all the old material. Warn your client ahead of time that the wood may not come out perfect. You don't know what's there until work begins.

A layer of varnish, lacquer or stain is usually easier to remove than several layers of paint. The most difficult stripping job is removing paint from old cabinets or doors and then staining them. The wood has to be perfectly clean before applying stain. Even when you think the wood is spotless, little flecks of paint deep in the wood will show up as stain is applied.

Equipment and Materials for Stripping

Wear industrial-grade rubber gloves when stripping. Stripper burns skin on contact. Wash immediately if any stripper gets on your skin. Wear a long-sleeve shirt and something to cover your head. A spray sock makes a good hat. The idea is to protect yourself as much as possible.

Use the best stripper you can buy. Cheap paint stripper is available. But it takes two or three times the effort to get the job done.

You'll need several old brushes to apply the stripper and several clean one-gallon buckets for dipping them. The buckets have to be spotless. If there's any old paint in the bucket, the stripper will transfer it to the wood, which will make your job just that much more difficult.

You'll also need two clean putty blades (2" and 3" width), several pieces of medium to fine steel wool, and lots of rags.

How to Strip

Start by reading the directions on the can. That's just common sense. Observe the manufacturer's safety precautions. One of these is sure to recommend thorough ventilation.

Apply the stripper to the surface with an old brush. Use a liberal amount of stripper, covering all areas thoroughly. Don't brush the stripper back and forth. This reduces its effectiveness. Just dab on a thick coat and let it work for a few minutes. If it starts to dry or evaporate too quickly, apply a little more.

After a minute or two, use a clean putty blade to scrape off the sludge. Be careful to avoid gouging the wood with the blade. Dump the sludge into an old can as you work. Then check the surface. If it's not clean, repeat the process. When you're satisfied with the condition of the surface, use T.S.P. and steel wool to scrub it clean. Then rinse with water.

At this point, you've done about all you can do with the stripper. Additional sanding and scrubbing with T.S.P. may be needed in some stubborn areas. You can still use wood bleach to clean and lighten the wood. Wood bleach is available at many paint stores. Follow the directions.

If you're going to spray lacquer on a surface that's been stripped, wash it thoroughly first with lacquer thinner.

General Painting Tips

Here are some painting tips that you may find worth the price of this book by themselves.

1) What's an effective hand cleaner? Painters don't have to spend hours each week washing their hands. And they don't have to suffer from constant skin irritation. Take a simple precaution before opening the first can of paint. Coat your hands and other exposed areas with Vaseline. This keeps paint from soaking into your skin. Any paint that does get on you will wash off easily and completely. When you stop for lunch, wash your hands and then reapply Vaseline.

Many painters wear cotton gloves. That will seem clumsy and awkward at first. But even soft gloves can protect your skin from constant abrasion. Replace the gloves when they're stiff with dried paint.

For the final cleanup of the day, use thinner and a commercial hand cleaner. For a thorough cleaning of resistant stains, scrub with a soft bristle brush and baking soda. You'll be amazed at the results.

To protect your hands from stain, use Vaseline and industrial-grade rubber gloves. Oil-based stain will eat through kitchen-type latex gloves in a few minutes. If you want to use Vaseline alone for protection without rubber gloves, apply it heavily. When you notice the stain starting to dry and cake on your hands, clean up with thinner and put on more Vaseline. This takes time. But it's much better than walking around with stained hands for several days.

2) What's better, oil or water-based enamel? Any experienced painter can give you a 30-minute argument on both sides of this question. Generally speaking, both are good and neither is clearly superior. But more important, the argument misses the point. The real key to a durable paint job is proper preparation and good application.

3) Do you need expensive paint to get a good paint job? No, proper preparation and good application technique are the keys to a quality, long lasting paint job. But avoid using the cheapest paints. They're generally inferior in quality.

4) Will thinning the paint save money? Not if it means having to apply an additional coat. Your highest cost is labor, usually 80% of the job. Paint is usually less than 20% of the job cost. Excessive thinning is a waste of time and material. Your aim is to get good coverage, get the job done and get on to the next job. The cost of one more gallon of paint is nothing compared to the value of the contract and the good will of your client.

5) If I only had one brush, what should it be? If I were using oil-base paints, I'd pick a 2½" straight-edge natural bristle. For water-based paints, select a 2½" straight-edge nylon/polyester bristle. These two brushes will get you through most jobs.

Chapter 9

Planning for Your Company

There are a lot of reasons to go into business for yourself. Making more money is only one. Few wage earners can earn enough in a lifetime to become financially independent. But many business owners can and do.

I'm not saying that running a paint contracting company is sure to make you rich. But I know many paint contractors with limited educations and no special qualifications who have done very well.

It's relatively easy for a paint contractor to take on a lot of business and generate a lot of income. Many new paint contractors are surprised to discover how much money they took in during their first year. But they are usually just as surprised to discover how little of that money stayed in the company as profit. Understanding why so little of that money stays in the business, and what you can do to keep more, is the subject of this chapter.

Everyone going into the paint contracting business should know the difference between *gross income* and *net income*. Gross income is the same as sales. It's what your clients pay for the services you perform. Net income (or profit) is what's left after all bills are paid (including a reasonable salary for the owner-contractor). Here's an example:

During a six-week period, Rembrandt Painting Company finished four jobs and collected in full for all four:

Jones job	-	$7,500
Brown job	-	6,100
Smith job	-	8,000
Thomas job	-	9,500
Total (gross income)		$31,100

Now, let's look at expenses for that six-week period:

Labor	-	$12,000
Materials	-	7,200
Job management	-	2,000
Insurance	-	150
Rent	-	1,500
Telephone	-	600
Truck	-	1,200
Promotion	-	2,400
Office supplies	-	300
Miscellaneous	-	1,650
Total expenses		$29,000

Subtracting total expenses from gross income leaves $2,100 as profit. That's a good profit if there aren't any hidden expenses still to be accounted for. Dividing the $31,100 gross into the $2,100 net, we get 6.75% profit. That's very close to our profit goal of 5% after tax.

But there's real danger lurking here. The income and expenses for a job don't come in at the same time. Sometimes you get paid before the bills come in. Usually the bills come in before you get paid. The danger is that high gross income gives you the feeling that you're making money. But when the bills come due, net income may be slim or negative, leaving you without enough cash to pay bills. That brings on a financial crisis — a cash shortage. And running out of money is the most common reason for failure of paint contracting companies.

A good understanding of the difference between gross income and net income (and a little financial planning) can help you avoid a financial crisis.

Unless you've had training in finance, the whole subject can be very confusing. Most paint contractors are good painters. They're also reasonably good at supervising and managing the business, and have learned to estimate costs and sell jobs. But financial planning and accounting are the last things most paint contractors want to learn. And it's usually the last thing they *do* learn. But learn it you must — if you want to avoid the most common financial blunders. Fortunately, learning a few simple principles will take you a long way.

Financial Planning

The term *financial planning* means anticipating where the money is coming from, where it's going, and making sure that there's enough available at all times to meet essential company needs.

Here's the most common error in handling money. The contractor collects the money on a completed job, pays the crew, and puts the rest in his personal bank account as the "profit" on the job. This is an absence of financial planning. It's a method of handling money based on hope. The contractor who does this *hopes* that he'll get enough money at the start of the next job to buy materials. He *hopes* that nothing goes wrong and that there are no delays on the next job. He *hopes* that his crew will be willing to keep on working until he collects some more money.

Here's another example of poor financial planning: Joe's Painting Company lands a big job. It's about six weeks of work. The contract price is $15,000. This is good work and should bring in a nice profit for the company. The agreement provides for payment of 10% ($1,500) to start the job. The remainder is paid in installments: 25% after the exterior is finished, 25% after all interior staining is done, 25% on completion of all interior work, and the final 15% when all touch-up is done.

Fix this fact clearly in your mind right at the beginning. Gross income from this job will be $15,000. But net income, what the company will make, will be about 5% of that, or $750, *if* everything goes according to plan. Remember that the net income is not Joe's salary. His salary is figured into the expense of the job. Net income is what the *company* will make. Any extra expenses or overtime work will reduce net income dollar for dollar. So there's a real risk of loss if the job starts to turn bad.

But Joe is optimistic. He plans to do good work, finish on time, and convince the general contractor he's working for that Joe's Painting is a capable, well-run, efficient operation.

Joe is a good painter. He works hard and knows the painting business inside and out. But he's never been very good at handling money. He usually works from job to job, leaving little money in the company checking account. If there's any money available, he takes it as personal salary. The result? There just isn't enough cash in the bank to carry a job this size.

When the initial 10% payment comes in, Joe sees only the $1,500. He doesn't realize that this is probably twice as much as he stands to make on the job — even if everything goes according to plan. His profit will be about $750. But already he has this check for $1,500. He thinks of that $1,500 as money made for landing the job, not as seed money to buy paint and pay painters. He's confused gross income (the first $1,500 payment) and potential net income (the $750 he may make).

With $1,500 in the company checking account, Joe starts to think about that new pickup truck he's been wanting to buy for the business. A painting company handling $15,000 jobs should have a shiny new truck, he reasons. He takes $500 of the initial payment and uses it to make a down payment on the pickup. That leaves him with just $1,000 to buy paint and pay his crew until the exterior is complete and he gets the next payment. It will be real close, he figures, but by putting in a few extra hours, he can just make it.

A few days into the job, Joe realizes that he needs a lift to reach several of the balconies on the top floor. Lift rental will come to $200. Plus, he underestimated the paint by 20 gallons. He'll need another $200 to buy the paint needed to finish the exterior.

So Joe is now $400 over his original estimate and only a few days into the job. That's bad. But another problem is worse. Right now, there's just enough cash to pay for materials needed to finish the exterior. There's nowhere near enough cash available to meet payroll. Until he gets the next progress payment, he can't pay his painters. By working overtime, he can probably finish the exterior by payday. But that will boost labor cost by close to $300 over his estimate. The only alternative is to ask the crew to wait a few days for their money. He gives the crew the choice of working overtime or waiting until the exterior is finished to get paid. The crew reluctantly agrees to work overtime. But Joe is walking on thin ice now.

So where is Joe at this point? He's digging his financial grave. He's spent an extra $400 on paint and equipment. Labor cost is running $300 over budget. Of the original $750 profit potential, $700 is gone. His crew is beginning to think about what they'll do when Joe's Painting goes belly up. The only embarrassing thing Joe hasn't had to do yet is ask his general contractor for an advance. But that probably isn't far off. After all, this is just the first week of the job!

If this seems like a ridiculous example, you haven't been around the painting business very long. It's exactly the type of thing that happens all the time. The primary reason? Poor financial planning.

Joe will probably stumble through this job, with more than a few headaches and lots of worry. And when he finally does collect that last payment (and the $50 profit), he won't have any more cash in the bank than when he started the last job. The whole cycle will start again when he signs the next big contract.

Joe isn't unique. There are many painting contractors who handle their business this way week after week, year after year. It's called being in a permanent cash crisis. It forces contractors to take work they would rather pass up. It forces contractors to price their work below cost so they keep the money coming in — even though doing that makes the crisis worse. It forces contractors to waste time, material, manpower and scarce cash resources because they have no other choice.

What's the alternative? Easy. Have a plan. Plan right from the start to have enough capital. No matter how limited your cash, there's some level of business your bank account can support. Stay within that level of business. Take only potentially profitable work. Keep a cash cushion as insurance against a rainy day. Don't siphon off all spare cash as soon as it hits the company bank account. Don't even take all the profits out when the profits are finally earned. Leave enough cash in your business. Working capital is the lifeblood of every business. When cash runs out, the heart of your business stops beating. Don't let that happen.

Get Started with Your Planning

If you don't understand bookkeeping, get some expert help. A small contractor with only two or three employees needs only a few hours of an accountant's time every month or two. Get a payroll service and a bookkeeping service to process payroll and keep your books. The cost will be about $50 a month. Many banks offer this service at a modest cost to their account holders. If your bank doesn't have a payroll service and an accounting service, get their recommendation on someone who does. If all else fails, look under "Data Processing Services" in the Yellow Pages of the phone book.

Unless you're already a computer whiz, don't even consider doing this work on your own computer. You have better things to do with your time and money.

If you don't already have a set of accounting books, ask an accountant to set up some books for your company. You don't need a C.P.A., just a competent accountant. If that accountant already has several paint contractor clients, that's a plus for you.

Once the books are set up and you have a routine for handling receivables, payables and your general ledger, you'll need the help of an accountant only occasionally. The accountant should review your accounts every three months and prepare tax returns, closing entries and a depreciation schedule at year end. Don't try to handle your company's state and federal income taxes yourself. It takes an expert. Your accountant should be current on tax regulations and will make out the company returns for your signature.

Of course, your accountant doesn't handle company money. He can advise, help with your taxes, and help anticipate problems. But financial planning is still your responsibility. You, as the owner of the business, are ultimately responsible for what happens to company money.

A System of Accounts
In a larger company, each operating division has a budget and is required to operate within that budget. For example, the marketing department would have an annual budget of a certain number of dollars. Unless the budget is revised, marketing has to stay within that budget.

Drawing up a formal budget isn't necessary or even desirable in a smaller company. But you can get many of the advantages of budgeting with a simple system of multiple accounts. To do this, you'll have to open six new accounts at your bank, one for each of the six functions I'm going to describe. We'll call these your *operating* accounts. Your bank may impose a small monthly charge to maintain each of these accounts. And you'll need printed checks for each account. But the benefits will make the additional cost worthwhile.

The advantage of this system of multiple accounts is that the bank does the record-keeping for you. You'll get a monthly statement for each operating account, showing all the transactions for that account. Of course, you have to reconcile each of the six account statements. But balancing six accounts with a few transactions is usually easier than balancing one account with six times as many transactions.

Here's how my system works. All money you receive from clients goes straight into a single account. We'll call this the *income* account. Make sure every check and all cash is deposited to this income account. Write checks against the income account as necessary. But money from the income account goes only to the operating accounts, never anywhere else. And how much is disbursed to each of the six accounts is always a set percentage of the total paid out of the account.

Every time you take money out of the income account, write six checks, one to each operating account. The percentage each receives is up to you. But the percentages shouldn't change from month to month. Set a percentage for each account and stick to those figures as long as possible.

This system forces you to divide all income into six pockets, making it less likely that you'll blow too much in one area and leave nothing for others.

Here are the six operating areas that have a claim on all income.

1) Reserve account— This is money that's set aside for an emergency. It's the backbone of your company. At first it will be difficult to set this money aside. But consider your reserve contribution a bill that has to be paid. This account system is a forced savings plan. The reserve account is where the savings go. After you've made regular payments to the reserve account for several months, the growing balance will make it easier to keep making more contributions.

When a big contract comes along, when you have the chance to pick up some much-needed equipment at fire sale prices, here's where you'll find the cash to swing the deal.

How much goes into the reserve account? That's up to you. 10% wouldn't be too much if you can swing it. But 5% or less is probably a more reasonable expectation.

Your reserve should be in an interest-bearing account. You won't have many withdrawals from this account. Have your bank describe the many types of accounts available.

2) *Promotion account*— Write checks on this account for all promotion expenses: newspaper ads, leaflets, stamps, ads in the Yellow Pages, etc.

Most contractors spend too little on promotion. Some spend nothing at all promoting their services. That's a mistake. Without promotion, no one knows you're there. Don't cheat your promotion account, especially when work is slow. This is exactly the time to promote, promote, promote!

How much do you allow for promotion? Again, that's up to you. It would be nice to spend 10 cents of every sales dollar on promotion. Do it if you can. More likely you'll spend something less than that. In any event, don't go less than 4 or 5%.

3) *Payroll account*— Payroll is your biggest expense. At least 50 cents of every sales dollar will go for payroll. Allocating one-half of all receipts to meet payroll should make it more likely that you'll have the money there when it's needed.

4) *Material account*— Draw on this account to buy paint, thinner, drop cloths, rags, hand tools, minor equipment, and all the things needed to do the job.

There's no way to skimp on this account. Materials will cost at least 15% of gross income. For some contractors it will be 20% to 25% of income.

5) *Overhead account*— Write checks on this account for all the expenses incurred to keep your business going: rent, phone, insurance, answering service, office supplies, licenses and utilities. Also lump into this account any expenses that don't fit elsewhere.

If you're tempted to buy a new copy machine, sure, go right ahead — as long as there's enough cash in the overhead account. If not, your old copier will have to hold together for at least one more month. Don't even consider borrowing from the material account or the payroll account. That's not allowed!

Overhead (as defined here to exclude office payroll and promotion) should be less than 10% of income. For most painting contractors it will be under 5%.

6) *Building account*— This account is for building your company's productivity. Some day you may use money from this account as a down payment on an office for your company. Until then, dip into this account to buy a new spray gun, new van or truck, or some scaffolding.

Your building account might get as much as 5% of gross. But 2½% is probably a more realistic figure.

Having these six accounts is like wearing a suit with a dozen pockets. If you stuff a little cash into each when cash is available, you're almost sure to have some change in one pocket in an emergency. And in the meantime, you've forced yourself to budget for the key operating areas of your business.

This system of multiple bank accounts should budget expenses in each operating area of your business. But once you have this ability, keeping a single bank account with 6 expense categories will be more practical.

Setting Up Your Accounts

How do you decide what percentage of income to allocate to each account? That's easy. Just do a little analysis of how your money has been spent.

It's a little harder if you're just starting out in business. Your figures will have to be estimates. Make your expense estimates a little on the high side just to be safe.

If you've been in business for awhile, review company check stubs for the last couple of months. Break all expenses into one of the four major categories (promotion, payroll, materials and overhead expenses), then project future expenses based on those figures. Figure 9-1 shows a summary of projected monthly expense totals for a small painting company. If expenses for a month will total $8,800, dividing by four gives us an average weekly expense of $2,200.

Next, we'll figure gross income over a similar period. Let's say that for the last eight weeks, total income was $24,000. Divide that by eight to find the average weekly gross of $3,000. See Figure 9-2.

Now you have two key figures: average weekly expenses and average weekly income. If income is less than expenses, you're in trouble. You better

CATEGORY		MONTHLY COST
PROMOTION	YELLOW PAGES	400.00
	NEWSPAPER AD	200.00
	FLYERS	100.00
	MAILING COST	100.00
		$800.00
PAYROLL	ONE JOURNEYMAN	1800.00
	ONE APPRENTICE	1000.00
	OWNER'S PAY	2800.00
		$5600.00
MATERIALS	PAINT	
	THINNER	
	BRUSHES	
	ROLLERS	
	TOOLS	
	EQUIPMENT RENTAL	
	PREP MATERIALS	$1400.00
OVERHEAD	RENT	
	PHONE	
	UTILITIES	
	OFFICE SUPPLIES	
	XEROXING	$1000.00
	TOTAL	$8800.00

Cost estimates to operate one month
Figure 9-1

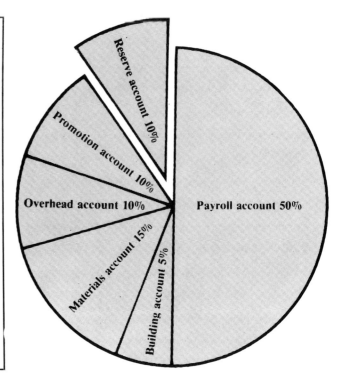

Determining weekly gross income
Figure 9-2

Percentage of income in each expense account
Figure 9-3

figure out some way to cut expenses and increase income.

Now we can figure what percentage of income falls into each of the six operating categories.

Average weekly gross income = $3,000.00
Percentage breakdown:

Reserve account	10%	$ 300.00
Promotion account	10%	300.00
Payroll account	50%	1,500.00
Materials account	15%	450.00
Overhead account	10%	300.00
Building account	5%	150.00
Total	100%	$3,000.00

These percentages are just an example. The figures you develop yourself will be different.

Every company is unique. Figure 9-3 shows how the income pie is divided in our example.

With these figures, the company now has a simple system that lets the owner anticipate where each revenue dollar will go. These percentages also provide a goal. Notice that the building account and the reserve account were not part of the original expense summary shown in Figure 9-1. They're included now because your goal is to build reserves and accumulate some cash in the building account.

Of course, income will change from week to week and from month to month. If income is way down, contributions to all accounts will be down. But the percentages should be the same.

If this system of allocation seems like extra work, you're right. But it also enforces a kind of financial discipline on your company. It's the same kind of discipline that some of the largest and most successful companies use to control expenses. If it saves you from a major financial crisis, it's well worth the trouble. There's no denying that this allocation system will:

1) Create a budget for each operating area of your business
2) Ensure that all of your expenses are covered
3) Create some reserves for use in an emergency.

Financial Planning Day

One day each week should be your financial planning day. That's when you take an hour or two to handle financial records, distribute income to the operating accounts, and plan for the coming week. The planning day should coincide with your financial "cutoff" day. All receipts that come in before that time fall into the previous week. Receipts that come in late go into the next week.

On your planning day, add up all the money collected for that week. Deposit it in your income account and then distribute income to your operating accounts according to the percentage breakdown you follow.

This is also a good time to estimate income and expenses for the coming two weeks. Use a chart like Figure 9-4 to enter receipts when you expect to receive them and expenses when they are expected to come due. Summarizing expected income and expenses for each week helps you anticipate the ups and downs in your bank balance. In our example, Thursday is the cutoff day, so each weekly period begins with Friday.

Planning in a Nutshell

A little bit of financial planning knowledge is important. But what I've explained here can go a long way. Going deeper into this subject would probably be a waste of your time. Painting contractors don't have to be expert financial planners.

I'm not going to cover bookkeeping or office management. Each of these is a major topic that can fill an entire book. If you're interested, there's a list of such books and other books for contractors at the end of this volume.

But I want to cover two more topics before we leave the financial field. The first is the ledgers you should keep. The second concerns collection of past due accounts.

Payment Ledgers

Most of the larger jobs you handle will have a payment schedule in the contract, rather than a single lump sum payment. Here's an example:

Name: Jones job
Type of job: Paint exterior of house
Price: $3,500
Payment schedule:

10% down at start of job.............	$ 350.00
20% after all prep work completed....	700.00
25% after all exterior painted........	875.00
25% after all trim work completed....	875.00
20% upon completion of all work including touch-up....................	700.00
	$3,500.00

It's important to keep a payment ledger any time you receive progress payments. These are admissible in court to show what has been paid. But more important, it's your record of what work was done, the balance due, and the date the money was received. Figure 9-5 shows the ledger for the Jones job. Your ledger could be an inexpensive notebook with 8½" by 11" pages. For each job that has multiple payments, start a ledger sheet and record all payment activity for that job.

A set of ledgers like this makes it easy to anticipate when income will be received. Looking through the ledgers will show at a glance when payments can be expected for each job in progress.

Collecting Your Money

At the end of every good job is a check for payment in full. If the work you do is professional quality and done according to your contract with the client, you should have no problem collecting. Unfortunately, there are exceptions.

Some people just don't like to pay on time, even if they have the money and have no legitimate reason to delay payment. Other clients feel like you haven't earned the final payment until they've picked your work to death. They always have "just one more thing" to touch up.

Many general contractors don't manage their money very well. If job costs are running over budget, the general contractor may run out of money before the job is finished. That won't affect the excavation sub or the framing sub. They've been paid long ago. But the painting sub is one of the last on the job. If the project becomes a loser for the general contractor, it's the painting sub who's going to have the hardest time getting paid.

Date	Day	Income	Expenses
16	Friday		$1400
17	Saturday		
18	Sunday		
19	Monday		$250
20	Tuesday	$900	
21	Wednesday		$825
22	Thursday	$2000	
	Totals	$2900	$2475

Date	Day	Income	Expenses
23	Friday	$700	$1400
24	Saturday		
25	Sunday		
26	Monday		$900
27	Tuesday	$500	
28	Wednesday		$350
29	Thursday	$2700	
	Totals	$3900	$2650

Projected income and expenses
Figure 9-4

Name: **JONES JOB** Contract price: $3,500.00

Date payment received	Description	Amount received	Balance due
21 Oct	10% DOWN FOR START OF JOB	350.00	3,150.00
24 Oct	20% FOR PREP WORK	700.00	2,450.00
29 Oct	25% FOR EXTERIOR PAINTING	875.00	1,575.00
1 Nov	25% FOR TRIM WORK	875.00	700.00
2 Nov	20% TOUCH-UP AND COMPLETION	700.00	—0—

Payment ledger sheet
Figure 9-5

Job Financial Summary

Date: 9/12/84

Job name: JONES

Original amount of contract: $4900.00

Changes or extra work added to original contract: BOY'S ROOM - $325.00

Total: $5225.00

Manhours: 240 × $13.00 PER HOUR = $3120.00

Materials: PAINT, SUPPLIES AND RENTAL COST - $650.00

Overhead: $1000.00

Total: $4770.00

Profit: $455.00

Additional notes: CUSTOMER WANTS EXTERIOR PAINTED IN TWO MONTHS. ROUGH ESTIMATE - $5600.00. PREPARE FINAL ESTIMATE IN TWO WEEKS.

Job financial summary
Figure 9-6

So how do you handle these collection problems? A good collection policy is the best way I know. Debtors pay creditors when it's easier to pay than to keep on stalling. You just have to make it easier to pay than delay. Do that by laying down some ironclad rules to follow:

Collection Policy

1) We base all jobs on a written contract with a specific payment schedule.

2) If a client misses a payment, work stops on that job until payment is made.

3) When a payment is 30 days late, we make a request for payment either in person or over the phone.

4) Every 10 days thereafter we make another request for payment until the bill is paid in full.

5) After 90 days, we file suit for the full amount due.

6) If we receive a bad check, we ask the client to make the check good by delivering a cashier's check, money order or cash to us.

But here's a collection policy for *you* to follow. *Never try to collect before the work and touch-up are finished.* This puts your professionalism in question. Do all of the work according to the payment schedule. *Then* collect your money.

Job Financial Summary

One final point before we leave the financial area. I recommend that you make up a financial summary for every job. See Figure 9-6. This form summarizes some important information that you may want to refer to from time to time: gross receipts, the actual number of manhours expended, the actual cost of materials, your profit, and any comments that you want to preserve. Keep a file of these summaries in a three-ring binder so it's easy to refer to them as needed.

Chapter 10

Planning for Growth

Many paint contractors find that it's easy to increase business volume — up to a certain point. As your business grows, you just hire more painters, do more estimating and increase promotion. But at some point, many painting contractors come up against limits that make further growth either unwise or impossible.

Running one or two crews is relatively simple. If you're a good painter and are willing to do some supervision, you can handle five or six men without too much trouble. But there's a limit on how much work one person can supervise. Beyond that second crew, it's easy to lose control. Coordination becomes more difficult — and more important. Your skills as a painter become almost irrelevant. What counts is how good you are at supervision and business management.

With more than two or three crews, complexity grows. There's more of everything to handle. More employees to supervise, more customers to deal with, more money to collect from more sources, more problems to handle, and more payroll checks to issue. A five-crew paint contracting company isn't a small business that can be controlled easily by one person.

I'm not going to insist that every paint contracting company should keep growing so long as it's profitable to do so. For many contractors, five or six painters on staff will be enough. Only the contractor running the show can make that decision. But if you want to build a larger painting company, this chapter will outline the principles that I've found to be important. Follow my suggestions here and continued expansion may be much easier.

Grow Gradually to Avoid Overextension

First, recognize that there are real limits to how fast you can expand. Doing more work requires more equipment, more skilled and motivated painters, more capital, more streamlined and effective procedures, and more supervisors. Accumulating resources like these takes time. If your business is doubling in size every six months, there are bound to be severe growing pains.

My experience taught me that growth should be in small steps, followed by periods of consolidation. Avoid giant leaps forward. This is the most basic principle of expansion. It's also the one most often ignored by contractors. Violating this principle of gradual growth gets more contractors into more trouble than anything else. Overextension could be called the contractor's deadly disease. All too many catch it.

There are many ways to overextend yourself. All amount to the same thing — reaching beyond your grasp. It's an easy trap to fall into. The potential rewards are tantalizing.

Here's a typical example of overextension. Put yourself in the position of the contractor here and see if you could resist the temptation.

Your company consists of yourself and two other painters. It's a small operation. But it usually runs smoothly and you have a good reputation for finishing on time and doing quality work. Capital is a little short. You've got less than a thousand dollars in the bank right now. But, as long as there are no delays or problems, limited capital hasn't been a problem.

You've just started work on an office building. It should take about eight days to complete. While you're on that job, your answering machine takes a call from a client who asked for a bid on his home last week. You return the call. The client accepts your price, but he wants work to start this week so it's finished for his daughter's wedding reception. The client agrees to sign the contract if he has a commitment from you to start work by the end of the week. Because you have a reputation as a trustworthy and quality-conscious painter, he's willing to make a substantial advance payment.

You agree to take the job. It's hard to refuse profitable work like this. It's exactly the kind of work you want to have plenty of in the future.

Now, let's get practical for a minute. There's no way you can pull your crew off the office job you're on now. You're going to have to hire a new crew of painters and put them to work on the bride's home.

After a few calls, you locate three painters to take on the new job. Only one has worked for you before. And he wasn't very reliable. Now he's your lead painter on an important job!

The bride's job gets started on Friday. But you have to shuttle back and forth between two projects now. That makes you more of a supervisor than a painter. It's a strain, but everything goes O.K. the first day. At least there aren't any major upsets. But you notice that work seems to slow down on the office building while you're supervising the other job.

By the following Wednesday, the interior of the bride's job is nearly finished. But the exterior hasn't been started yet. That's bad news because it's raining. The crew can finish the interior, but exterior work will have to wait until the weather clears. Work on the interior of the office building can go on in the meantime.

Late Wednesday afternoon you get a call from a builder you worked for two weeks ago. He insists that you forgot to stain a wall cabinet in his spec home. He says the unfinished cabinet sticks out like a sore thumb. He can't close the sale until the cabinet is stained to match the rest of the kitchen cabinets. You agree to drive over to the job before nightfall.

You meet the builder at the spec home and discover that he's right. The wall cabinet isn't stained. But that cabinet wasn't hung, it wasn't even on the job when your crew was there! The builder agrees that you're probably right. But he points out that the cabinet is on the plans. That means you've been paid for staining a cabinet that didn't get stained. He insists that it has to be stained and finished by this time tomorrow. You agree to do the work. And, unfortunately, you're going to have to do it yourself. There's no one else you can trust to match the stain on the existing cabinets.

While you're working on the cabinets Thursday morning, you realize that there's not enough cash in the bank to pay both crews in full. Both expect to be paid tomorrow, but neither the bride's job nor the office building will be finished until next week. The cabinet in the spec home will be finished as promised. But you've already been paid for that. Where's that payroll money going to come from?

Friday morning dawns clear and warm. Perfect painting weather. Your first stop is the bride's home. The interior is finished and work has started on the exterior. The job should be finished Monday — in plenty of time for the reception. Your client is pleased. But the three-man crew spent all day Wednesday and most of Thursday on the interior — doing only about as much work as a two-man crew should handle in half the time. There goes the profit on that job.

You're pretty mad at those painters for dragging out the interior work. But that's just human nature. It was either that or take the day off without pay. The lead painter has no experience as a supervisor. Under the circumstances, he did a reasonably good job. And there's no advantage to criticizing their work now. You still have to break the bad news that there isn't enough cash to pay for the past week's work.

The real problem is at the office job. It's a little behind the schedule you laid out, but nothing that can't be made up in a day or two. But when you explain that half of payroll has to wait until next week, both painters quit. Neither is willing to work for you until their pay is brought up to date. Of course, you can finish the job yourself by working nights and weekends, but that will delay payment even longer. And you have to eat, too!

At just about this time you get a call from an architect's office. He has a custom job for you like you wouldn't believe. You'd make more on this than on both the other jobs together. But it has to be started now. Because of the mess you've gotten yourself into, you'll have to turn it down. He'll probably call somebody else next time.

Expansion is Limited by Resources
In the example above, you can see two missing ingredients. Obviously, the contractor was short on cash. But the lack of trained painters and supervisors was a more serious problem. Ideally, there should be no more than five employees for every supervisor. Beyond that, the supervisor loses effectiveness. Thus, if you hire on five new employees and don't have anyone trained to supervise them, you'll either have five unsupervised employees or ten under-supervised employees and one exhausted, partially effective supervisor. Either way, poor production will be the result. This leads to missed schedules and lower profits. That's why many painting companies double in size and workload yet don't earn a penny more for their effort.

Always include competent supervision in your expansion plan.

Schedule jobs too close together and you have jobs overlapping each other. That may make it necessary to split crews or move them from one job to another prematurely. When crew members are on a merry-go-round, production suffers.

Taking on more work and employees than your bank account can handle is deadly. But painting contractors do this all the time — hoping to collect money as the work goes on. Sometimes that works. But a little bad luck and there's no margin to fall back on. Once cash gets tight, the only important consideration is bringing in a few dollars. That clouds judgement and makes profits less likely. Once you're behind the eight-ball, it's very easy to stay there. Eventually you may go broke.

The solution? Stay within the bounds of your bank account.

Taking Gradient Steps
Expansion by gradient steps is taking one small step at a time. Each step you take should be only as large as you can comfortably handle. Then pause and consolidate the gains before reaching out for the next step.

When business is up 40% or 50%, your working capital has to be 40% or 50% greater than it was. If it isn't, you're courting danger. All expansion in work load has to be accompanied by expansion in the capital available to you. That's why I recommended that you set aside a percentage of your weekly gross income for expansion.

Nearly all expansion will require additional employees: painters, apprentices, and helpers. When your present staff is producing at peak capacity, taking on more work means taking on new employees. You can usually find journeymen, apprentices and helpers who can take instructions. But it's harder to find good supervisors who can give instructions.

Adding a painter, apprentice or helper is a small step for most painting companies. You can add a man or two almost any time without stretching resources too thin. But adding a supervisor is a big step.

Here's how to reduce the size of that step. Instead of adding supervisors when work gets plentiful, make supervisors out of your journeymen. Train them to work independently. Show them how you do business. Prepare them for the day when they have to step in and take charge of a crew or two. That's promotion from within. It's a smaller step. And it's going to create a smoother transition than hiring someone new to supervise. Expansion of your workload is limited by the resources you have available — finances, equipment and trained personnel. Never take on work that's beyond your ability to produce. If the job is too large, if it takes more experience or cash than you have available, then turn it down. Taking on work you can't handle is a sure way to lose money, customers and reputation.

The most basic rule of expansion is this: Never take on work that stretches your resources beyond reasonable limits or forces a compromise in your professional standards. Increase your workload. But do it in small increments. Each increase in workload will present new problems and challenges. Expanding in small steps keeps you in control.

If this sounds basic and simple, that's because it is. Breaking this rule is very tempting to all contractors. But painters who ignore this rule frequently wind up in trouble. Many conclude that expansion is impossible and give up any hope of running a larger company.

Make expansion a gradual process. That means it's a slow process. Your company won't become a giant corporation overnight. But neither will it collapse before dawn tomorrow.

Increasing Profits, Not Volume

Some contractors prefer to remain small. They either can't or don't want to have responsibility for running a larger company. Instead of being the biggest painter around, they're content to become the most profitable. They concentrate on improving the quality of their work and the quality of their clientele. Their long-term goal is to pick and choose among the better-paying jobs. That's an excellent way to increase income without increasing your company's size. It could be a good model for your painting company if you're uncomfortable running a company with dozens of employees.

Find Your Level of Competency

Every company has a level of competency. Yours is determined largely by your management skills and the talent of your supervisors. If you go beyond this level of competency, you may work harder and longer hours, but you won't necessarily earn any more money. Business is likely to become chaotic and out of control. At this point, you've got to reduce volume to a level where you can regain control. Expand again when you've found solutions to the problems you encountered.

Look for Profitability

Any time your company goes through a period of growth, profitability becomes a key issue. Keep track of operating profits. If increases in volume don't increase profits, and if it doesn't look like the new work will become more profitable, the extra revenue isn't doing your company any good. Stick to painting apartments if that's profitable. Drop the tract housing market if the money isn't there. Reinforce what works. Abandon what doesn't.

Learn to Say "No"

The ability to say "no," to turn down work, is just as important as being able to sell jobs. All contractors are flattered to have work shoved their way. Some never learn to refuse it. But being overloaded and unable to complete work on schedule does neither you nor your client a favor.

If you're completely booked up carrying two jobs, don't try to squeeze in a third. You'll just wind up with three unhappy clients. People respect honesty. They appreciate being told, in all candor, that you're too busy right now. They won't forget you. In fact, they'll probably put a premium value on your service next time.

There will always be more work available. I know contractors who worry about where their next job will come from, even when they're booked solid for the next two months. This isn't necessary. There's always work available. If you're having a problem finding work, you've been neglecting the promotion I recommend.

The Last Word

Here's the final word on expansion. Nothing in this universe stays the same. Everything changes. Change isn't always perceptible to those in the middle of it. But it always goes on. This is true of companies, also. Your company will never stay the same. It will always change, and there are only two ways it can change. It either expands or contracts. Up or down.

It's unusual for a company to survive without growing. Expansion creates new opportunities for your employees. It creates a feeling of success and enthusiasm for everyone in the company. Stagnation breeds a feeling of helplessness. It closes off opportunities for promotion within the company. No one wants to work for a company that's stagnant. That's why it's in your interest to see the company grow, even if accepting new opportunities does create new challenges.

Paint Shopping List

Brand	Name	Code Number	Flat	Enamel	Quantity

Common Materials Estimate

Item	Type	Quantity	Cost
Sandpaper			
Sandpaper			
Spackle			
Patching material			
Putty			
Caulk			
Paint thinner			
Lacquer thinner			
Rags			
Roller			
Roller			
Tape			
Plastic drops			
		Total	

Equipment Estimate

Item	Description	Cost

Estimate Summary

Item	Price
Painting and materials	
Common materials	
Equipment	
Subtotal	
Overhead and profit - _____%	
Total	

Planning for Growth

Payment Ledger Sheet

Name:		Contract price:	
Date payment received	Description	Amount received	Balance due

Job Financial Summary

Date:_____

Job name:_____

Original amount of contract:_____

Changes or extra work
added to original contract:_____

                                    Total:_____

Manhours:_____

Materials:_____

Overhead:_____

                                    Total:_____

                                    Profit:_____

Additional notes:

Index

A

Accidents 26-27, 145
Accountant 201
Accounting books 201
Acid-etching 165
Acoustic ceilings . . 174, 178, 179
Acoustical ceilings, manhours . .
. . . . 117
Acoustical surfaces . 152, 164-165
Acrylic paint 166
Adhesion 170
Admissible evidence 34
Ads
   classified 47, 49
   display 49
   newspaper 47, 49
   Yellow Pages 49
Advertising
   effective 45, 47, 49-50
   false 45
Advertising copy 41, 45
Agreements, verbal 33, 75
Airless spray guns 178, 195
Alcohol 194
Alcoholic beverages 30
Alkali 150
Alkyd resin 150
Aluminum 165
Ammonia 170
Apartments
   estimating 77
   list prices 62
Appearance 29
Apprentice 26
Architects 29, 142, 150
Associated General Contractors .
. . . . 49

B

Barn walls 152
Base prices 62
Baseboards 154, 174, 175
Basic tools 148
B.C.P. 166
Beam details 86
Beams 83-84, 86, 154
Bid
   contract 77
   courtesy 59
   cover letter 76
   lowest 78
   price 59, 69, 71
   submitting 75
   winning 61
Bidder, lowest responsible 58

Bidding
   for interior designers 62
   volume 59, 61
Biographical data 52
Blistering 150, 169, 171
Block fillers 152
Blueprint
   changes 78, 80
   reading 78
   take-off checklist 81
Blueprints
   beam details 84, 86
   cabinet details 84-85
   changes 78, 80
   details 83
   door schedule 88-89
   elevations 83
   finish schedule 83-84, 87
   floor plan 82
   paneling details 84, 86
   section views 83, 84
   specifications 89-92
   take-off checklist 81
   window schedule 87-88
Bonus 23
Bookcases
   lacquering 122
   manhours . . 113, 116, 121-122
   on blueprints 84
   painting 113
   spraying 116
   staining 68, 121
Bookkeeping service 200
Booze 30
Boxing the paint 173
Bristle brushes 148, 162
Brush
   drag 184
   cleaning 154-156
   complaints 156
   sizes 154
Brush comb 154, 155
Brushes . . 148, 149, 152, 154-156
Brushing
   enamel 179-191
   flat paint 173-176
   lacquer 196
   stain 194
   stripper 196
   varnish 195
Buckets . . . 22, 145, 149, 154, 157
Budgeting 201
Building account 202, 204
Bulk rate mail 47, 53-54

C

Cabinets
   brush size for 154

   details 85
   lacquering 122
   manhours . . . 67, 113, 116, 120
   on blueprints 84-85
   painting 113, 190-192
   spraying 116
   staining 120
Cage handle 149
Calcimine 164
Capital 8
Capital, working 211
Cards, business 41
Carpet, protecting 174-175
Carved doors 184-185
Catalyzed enamel 152
Caulking cracks 163
Caulking gun 158
Ceilings
   acoustical 117
   brush size for 154
   checklist 123
   manhours . . . 67, 113, 117, 121
   smooth 113, 116, 117
   staining 121
   textured 113, 117
   tongue and groove . . . 113, 117
Cement 152
Cement base paint 152
Central files 47
Chain of command 25-26
Chalking 164
Changes, blueprint 78, 80
Chasing the paint 173
Cheats 59-60
Children, working around . . . 145
Chrome 165
Cinder block 152
Classified ads 47, 49
Cleaning brushes 154-156
Cleanup, jobsite 147, 150
Clear base paints 143
Client
   financial trouble 31
   relations 28
   satisfying 142-143
Clientele 59-60, 61
Clients
   being selective 60
   difficult 62
   expectations 63
   upset 60
Clogging, spray tips 178
Clothing 20
Collecting 31, 77, 205, 208
Collection policy 208
Color
   changes 64
   in contract 77
   schedules . . . 142-143, 144, 150
   selection 142
Commercial buildings 77-78
Commodity grade work 28

Common materials estimate . . . .
. . . . 69-70, 72
Communication 22-23, 31
Company
   goals 7-8, 18
   logo 41, 45
   meetings 13-14
   name 40
   policy 15, 20
Competition 28
Competitive conditions 62
Completing tasks 13
Concrete floors 165
Concrete, manhours
   patios 118
   porches 115
   walls, smooth 119
Concrete, paint for 152
Condominiums, list prices 62
Conflicts 31
Consistency
   enamel 180-181
   flat paint 173
Construction etiquette 32
Contract
   acceptance 75
   and trust 29
   completeness 31
   estimate 69
   rescission notice 37
   sample 35-37
   standard 77
   terms and conditions 36
Contractor license number . . . 42
Contractors
   and other professionals 29
   image 34
Coordinating colors 142
Copy, advertising 41, 45-46
Corner-cutting on specifications
. . . . 89
Cost estimates 203
Cost Records for Construction
Estimating 59
Costs
   actual 59
   equipment 71
   estimating 63, 65, 75
   labor 69
   manhours 65-68
   materials 69, 71
   overhead 71-72
   preparation 64
   tools 71
Coverage 92, 143
Cover letter for estimate 69
C.P.A. 201
Cracked paint 170
Craftsmanship 28
Crayon marks 173
Credit application
   on large jobs 33

219

sample 38
Crew time, estimating 63
Cross-grain cracking 169
Custom colors, cutting in ...176
Custom jobs
   painting 61-62, 65
   pricing 59, 65
   specifications for 92
   staining 192
Custom-mixed colors 143
Customer
   chiselers 33
   etiquette 32
   irate 31
   needs 28
   relations 28-30
   satisfaction 34
   service 29
   survey 42
Cutoff day, financial 205
Cutting in 174-176

D

Daily plan, foreman's 137
"Deals" 33
Decorators 142, 143, 150
Depreciation schedule 201
Details, blueprint 83-87
Disc sander 168, 169
Discoloration 169
Dishonesty, employee 27, 30
Door
   edges 182, 185
   jambs 184, 186-187, 188
   schedule 88-89
Door-to-door leaflets 50
Doors
   enameling ..181-184, 185, 188
   estimating 64, 87-89
   exterior checklist 132
   flush ....
   ...66, 112, 114, 116, 118, 120
   French ....
   ...66, 112, 114, 116, 118, 120
   interior checklist 124
   lacquering 122
   louver ....
   ...66, 112, 114, 116, 118, 120
   manhours 66, 68,
   112, 114, 116, 118, 120, 122
   paint selection 152, 154
   panel ....
   ...66, 112, 114, 116, 118, 120
   spraying 116, 118
   staining 60, 120
Double hung windows ..188, 190
Downspouts 152
Drinking on the job 30
Drop cloths...148, 149, 157, 182
Dry brush 174, 175
Dry rot 190
Dust mask 169
Duster 148

E

Efflorescence 164, 167

Elevations, blueprint 83
Employee
   disagreements 30
   relations 31
Employment application ..20-21
Enamel undercoat 181, 188
Enamels..150, 152, 154, 179-191
Epoxies 150, 152
Equipment 148-149
Equipment estimate 71, 74
Error, margin for 61
Estimate
   delivering 61
   sample 93
   snap 63
   summary 71, 74
   take-off form 69
Estimates
   as indicator 9-11
   pending file 75
   starting out 61
   time and material 60
Estimating
   apartments 77
   basics 57-58
   by square foot 92-93
   checklists 64, 123-134
   commercial buildings ...77-78
   custom jobs 59
   fee 58
   forms 69-75
   labor 65-68
   materials 64-65
   new construction 78
   packet 75
   preparation work 64
   rules 58-59
   stainwork 68-69
   starting out 61
   steps 60-61
   tips 63
Estimating Guide 111
Estimators
   novice 59
   successful 58
Etching 165
Excessive wear 156
Expansion
   continual 212
   controlled 209, 211-212
   gradient steps 211
   gradual 209
Expense estimates 202-203
Expenses 199
Experience 20
Extension handle, roller 177
Exterior
   doors 188
   jambs 188
   paint selection 150-151
   preparation 166-171
   windows 188, 190
Exterior checklists
   doors and jambs 132
   siding 130
   stucco 131
   trim 134
   windows and French doors 133
Exterior features, elevations ..83

Exterior manhours
   painting 114-115
   spray painting 118-119
   staining 120-121
Exterior take-off form ....
   ....96-97, 100-101
Extra touches 32-33
Extra work 69
Eyeball estimate 57

F

Face siding, smooth, manhours
  119-121
Family rooms, estimating 67
Fascia board, manhours 114, 118
Federal Unemployment
   Insurance 24
Fences 154
Fictitious business name state-
   ment 41
Field supervisor ..16, 25-26, 139
Files 47
Financial
   planning 199-200
   planning day 205
   summary form 217
Fingering, brush 155, 156
Finish schedule 83-84, 87
Firing employees 26-27
"Fish eye" 194
"Fivers" 157
Fix-all 181
Flat paint
   brushing 173-176
   rolling 176-178
   spraying 178-179
   thinning 173
Floor plans 82
Floppy bristles 156
Flush doors, manhours 66,
   112, 114, 116, 118, 120, 122
Foreman ..26, 136-137, 142, 145
Foreman's plan 137
Freelance painter 7
French doors, manhours ....
   ....66, 112, 114, 120
French method, door jambs ...
  186-187
"Friends," beware of 33
Fumes 145
Fungicide 164

G

Galvanized iron 152
Galvanized metal 165, 171
General ledger 201
Glossy surfaces 163
Gloves 196, 197
Goals 7-8, 18
Grilles 152
Gross income ....
   .....9-10, 198-200, 202, 204
Gross receipts 208
Growth, controlled .209, 211-212
Gutters and downspouts, man-
   hours 115

Gutters, paint selection 152

H

Hand cleaner 196
Handrail, manhours ....
   ....112, 116, 120, 122
Hard-to-clean brushes 156
Hardware 152
Heel-hardened brushes 155
Helper 26
H.E.R.O. spray pump 178
High-gloss urethane 195
High-production painting ...135
Hiring 139
Holidays 173, 192
Hollywood method, door jambs
  186-187
Homeowner staining 192
Honesty 30
Hood, spray 178
Horizontal line, painting 174

I

Incentives 23
Income account 201
Indicators 9-11, 18
Initial payment 31
Ink stains 173
Insurance 24
Intercoat peeling 170
Interior
   door jambs 184-187
   doors 181-184
   paint selection 152-153
   trim 181-188
   windows 187-188
Interior checklists
   cabinets 127
   ceilings and walls 123
   doors and jambs 124
   staining and finishing.....129
   stripping 128
   trim 126
   windows and French doors 125
Interior designers, bidding for 62
Interior manhours
   lacquering 122
   painting 112-113
   spray painting 116-117
   staining 120-121
   stripping 119
   varnishing 121
Interior take-off form ....
   ....94-95, 98-99
Intoxication from fumes 145
Iron and steel 165

J

Job
   completion certificate 139, 140
   description 15, 16, 18, 33
   financial summary ...207, 208
   log 33-34
   scheduling board 141-142

Index

signs 50
work order 137, 138
Jobs
  scheduling 211
  sold 9-10
Journeyman 20, 26
Journeyman manhours ... 65-68

K

Kerosene 154
Key indicators 9-11
Knot sealer 169

L

Lacquer
  brushes 154
  retarder 195
  sanding sealer ... 194-195
  thinner 194-195
Lacquer-based stains 194
Lacquering, manhours .. 68, 122
Ladder safety 145-147
Ladders 145-148, 157
Lambswool rollers 156, 178
Lap marks 177
Latex
  caulk 158, 163, 166, 181
  enamel 180-181
  paint 150, 152, 154
Lath 162
Laying off 182-184
Leaflets 50-51
Legal costs 145
Letterhead, company 41-43
Letters
  general (newsletter) 47
  of recommendation 52
  personalized......... 46-47
Liability insurance 24-25
Liquid sandpaper 164
Living rooms, estimating ... 67
Logo, company 41, 45
Long john roller 176, 191
Long range plan, foreman's . 137
Loose paint 162
Louver doors, manhours 66,
  112, 114, 116, 118, 120, 122
Lowball painters 58

M

"M" pattern 176-177
Magazines, trade 49
Mailing
  lists 47, 53-54
  schedule 47
Mailings
  bulk 47, 53-54
  mass 47, 53-54
Major tools 148
Manhour tables
  exterior painting 114-115
  exterior spray painting 118-119
  how to use 111
  interior painting 112-113

interior spray painting 116-117
  lacquering 122
  staining........... 120-121
  stripping 119
  varnishing 121
Manhours for residential work
  cabinets 67-68
  ceilings and walls 67
  doors 66
  molding 66-67
  sash windows 66
  spray painting 67
Market research 40
Mask, spray 178
Masking tape 174-184
Masonry 152
Masonry paint 152
Material account 202, 204
Materials, common, estimate
  form 69-70, 72
Media kit 49
Medicare 24
Meeting checklist 14
Meetings 13-14
Metal surfaces, preparation . 171
Mildew 163-164
Mildewcide 164
Mineral coloring 152
Mineral spirits 150
Mixing paint 173
Moisture 168, 169
Moisture meter 168
Molding
  doors, painting 184
  brush selection 154
Molding, manhours ....
  ...... 66, 112, 114, 120, 122
Money requirements 8
Muntins 188, 189, 190
Muriatic acid 164, 167

N

Name, company 40
Name statement, fictitious
  business 41
National Construction Estimator
  111
Natural bristle brush 154
Neatness 22
Net income 198-200
New construction
  estimating 62, 65, 78-93
  staining........... 192-193
New wood preparation 169
Newsletter 47
Nickel 165
Nitpickers 143
Nonglazed brick 152
Novice painters 23
Nylon, polyester brushes ... 154
Nylon rollers 156

O

Office expenses 72, 75
Official record 33
Oil-base
  enamel 180, 197

paints 150, 152, 154
  primer 152
  stains 194
Opaque base paints 143
Open soffit, manhours 118
Operating
  accounts 201, 204
  expenses 8-9
  procedures 32
Operations statement 34, 39
Orange peel texture 184, 195
Organization 7-8, 22
Organization board ....
  .... 11-12, 14-15, 18
Other trades 136-137
Overextension 209-210
Overhead 71-72, 73, 202
Overhead account 202, 204
Owner's responsibilities ... 25-26
Oxidation 165

P

Paint
  bases 143
  estimator 57
  selection 150-152
  selection charts 150, 153
  shopping list 71, 73, 137
Painter's holiday .. 139, 173, 192
Painter's putty 181, 193
Painting & Decorating
  Contractors of America ... 111
Painting specs 92
Palm sanders 193
Panel doors, manhours 66,
  .. 112, 114, 116, 118, 120, 122
Panel doors, painting ... 184-185
Paneling details 84, 86
Paneling, manhours ....
  .... 116, 121, 122
Partial payment 31
Patching
  cracks 159-161
  holes................ 162
  material 159
Pattern rolling 176-177
Pay raise policy.......... 25
Payables 201
Payment
  ledgers 205, 206
  schedule 31, 205
  terms 77
Payroll
  account 202, 204
  service 200
Peeling paint . 165, 167, 169, 171
Pencil roller 176
Penetrating wood stains 194
Penetrol 180-181, 188
Percentage of income...... 204
Permanent cash crisis 200
Personalized letters 46-47
Personnel policy 15, 17
Pets, working around 145
Photographs 52
Pigment 164
Pigmented stains 194
Plan sheets 82
Plastic tarp 157

Policies
  company 15, 18, 32
  new customer 29
  statements 30-31
  written 31-32
Polyester
  paints 150, 152
  rollers 156
Polyurethane 195
Popcorn ceilings 117, 174
Positioning
  negative 46
  positive 45-46
Postage costs............ 53
Preparation, estimating ... 63-64
Price increases 61-62
Prices
  base 62
  list 62
  premium 68
  raising 59, 61
  standard 62
  time and materials 60
  undercutting 61
Pricing materials 65-66
Primer 188
Problem-solving........ 31, 34
Product quality standards ... 32
Production
  hours 9-10
  targets 137, 139
  work 62-63
Productive labor......... 65
Professionalism 29
Profit 7-8, 23,
  71, 75, 198-199, 200, 211-212
Projected income and expenses
  206
Promises 31
Promotion...................
  ..... 9-11, 41-42, 44-52, 54-56
Promotion account 202, 204
Promotional copy 45-46
Preparation
  exterior 166-171
  interior 158-166
  materials 158-159
  specific surfaces 163-166
  tools 158-159
Press book 50, 52
Prime coat 152
Primer 152
Primer application 165
Putty 148
Putty blades 148

Q

Qualified painters.......... 23
Quality
  appropriate 63
  standards 32

R

Rags . 147, 148, 149, 158-159, 193
Receivables 8, 201
Reconciling bank statements . 201
Red lead................ 152

221

Referrals 29, 52-53
Refinishing 192
Reliability 25
Remodeling, labor estimates . . 65
Repeat business 30
Reputation
    for craftsmanship 28
    for dishonesty 30
    for good service 29
    for quality 33
Reserve account . . . 201-202, 204
Residential estimate form . . . . . .
    . . . . 69, 70, 103-107
Residential work, manhours . . . .
    65-68
Resin 150
Resources 211
Respirator mask 196
Responsibility 25
Rewards 23
Right tools 152-157
Roller
    covers 149, 156-157
    grid 157, 176
    pan 176
    pole 149
Rolling
    enamel 184, 191-192
    flat paint 176-178
Ropey paint 188
Rough brick or block, manhours
    119
Rough sawn siding, manhours . .
    . . . . 119, 121
Rounded-off brushes 156
Rowdy behavior 30
Runner 163
Rust 165

S

Safety 64, 145-147
Safety equipment
    goggles 169
    mask 145, 178
Sales tools 50, 52
Sample estimate . 69, 93, 102-110
Sample specifications 90-91
Sandblast 163
Sandpaper . . . 148, 149, 159, 168
Sandpaper chart 193
Sash
    brushes 188
    glazing 190
Sash windows, manhours . . . . . .
    . . . . 66, 68, 112, 114, 120
Sash windows, painting . . . .
    . . . . 188, 189-190
Satisfying the client 142-143
Scaffolding 148
Schedules
    disruption of 31
    missing 32
    payment 31
    setting 32
    work schedules 29
Scheduling 135, 139, 211
Sealer 149, 173, 181, 194
Sealing wood 193-194
Section views 83, 84

Service
    customer 28
    selling 28
Shingle siding, manhours . . . . . .
    . . . . 119, 121
Shop area 157, 158
Shutters 152, 154
Side wear, brush 156
Signs 50
Single homes, list prices 62
Skills 20
Small shop 148-149
Smooth brick or block,
    manhours 119
Soaking brushes 156
Social Security 24
Solvent-thinned paint 150
Spackle . . 148, 149, 158, 159, 162
Spackle blade 162, 174
Spec houses, list prices 62
Specialty coatings 150
Specifications 89-92
Spinner 154
Spot priming 165
Spray
    hood 178
    mask 178
    sock 196
    tips 178, 179
Spray painting, manhours . . . . .
    . . . . 67, 116-119
Spraying
    acoustic ceilings 179
    equipment 178
    flat paint 178-179
    lacquer 195-196
    patterns 178-179, 180
    stain 194
    varnish 195
    urethane 195
Square foot, estimating by . 92-93
Stain-killing primer 159
Staining
    custom 192
    homeowner 192
    new construction 192-193
    new wood 194-195
    over existing finish 195
Staining equipment 193
Staining, manhours 120-121
Stainless steel 165
Stainwork, estimating 68-69
Stairs 152, 154, 165
Standard contract 33
Standard pricing 63
State Unemployment Insurance .
    24
Stationery, company 41-43
Steel and iron 165
Steps 152, 154, 165
Straight lines, painting 174
Strainer bags 182
Stripper 196
Stripping 196
Stripping equipment, materials .
    196
Stripping paint, manhours . . 119
Stucco 152, 166
Supervision 209-211
Supervisor's job description . . 16
Surface area, computing 79

Surface chalk 170
Survey form 44
Surveys 41-42, 44-45
Synthetic
    brushes 154
    lambskin 178
    roller covers . 156-157, 177-178

T

Tack cloth (rag) 181, 193
Tax returns 201
Taxes 24
Termites 190
Textured ceilings, manhours . . .
    . . . . 113, 117
Texturing patches 162, 163
Thinner 148, 180-181
Thinning
    enamel 180-181
    excessive 197
    flat paint 173
    lacquer 195
    stain 194
Timekeeping 139
Tinting undercoat 143
Tongue & groove ceilings,
    manhours 113, 117, 121
Tools
    owner supplied 148
    painter supplied 148
Tools, small 71
Total price 69
Touch-up 139, 143, 145
Touch-up
    list 143
    paint 143
Tract houses, list prices 62
Trade magazines 49
Transparent stains 194
Trim
    color 142
    decorative 84
    enameling 181-191
    paint selection 152
Trim, manhours . . . .
    . . . . 66, 112, 114, 120, 122
Troubleshooting 30
Trust 28-29
T.S.P. . . . .
    . . 159, 164, 165, 190, 195, 196
Turpentine 150
Twisted nylon bristles 156

U

Undercoat 143, 149
Underpricing 28
Universal tints 173, 193, 194
Urethane 68, 150, 152

V

Varnish 68, 150, 154, 195
Varnishing, manhours 121
Vaseline 193, 196-197
Ventilating fumes 145

Verbal agreements 33
Vertical straight lines 174
Vinyl wallcovering 166
Vinyl wash pretreatment 165

W

Wallcovering, prep for . . 165-166
Walls
    brush selection 154
    paint selection 152-153
    painting 173-178
    preparation 159-163
    spraying 178-179
Walls, manhours 67
Washing compound 159
Water-base
    enamel 181, 197
    paints 148, 154, 157
    undercoater 166
Water blasting 163, 167-168
Water damage
    estimating 64
    stains, painting 173
Water-thinned paint 150
Waterproofing 165
Weekly meeting checklist 14
Whites, painter's 22
Window
    brush selection 154
    paint selection 152
    schedule 87-88
Windows
    estimating 87-88
    exterior 188-190
    interior 187-188
Windows and French doors
checklists
    exterior 133
    interior 125
Windows, manhours . . . .
    . . . . 66, 112, 114, 120
Wire brush 154, 155, 162
Wire brushing 164
Wood
    bleach 196
    paint selection 151-153
    putty 193
    surfaces, wet 164
Wood decks and porches,
    manhours 115, 119, 121
Wood window screens,
    manhours 115
Work
    completed 9-10
    quality of 29
    schedule 135-136
    summary 137
Worker's Compensation
    Insurance 24
Working capital 8, 200
Written agreements 33
Wrought iron fence, manhours .
    . . . . 115, 118

Z

Zinc chromate 152
Zinc dust primer 171

Practical References for Builders

Painter's Handbook

Loaded with "how-to" information you'll use every day to get professional results on any job: the best way to prepare a surface for painting or repainting; selecting and using the right materials and tools (including airless spray); tips for repainting kitchens, bathrooms, cabinets, eaves and porches; how to match and blend colors; why coatings fail and what to do about it. Lists 30 profitable specialties in the painting business.
320 pages, 8½ x 11, $33.00

National Construction Estimator

Current building costs for residential, commercial, and industrial construction. Estimated prices for every common building material. Provides manhours, recommended crew, and gives the labor cost for installation. Includes a CD-ROM with an electronic version of the book with *National Estimator*, a stand-alone Windows™ estimating program, plus an interactive multimedia video that shows how to use the disk to compile construction cost estimates. **616 pages, 8½ x 11, $47.50. Revised annually**

National Painting Cost Estimator

A complete guide to estimating painting costs for just about any type of residential, commercial, or industrial painting, whether by brush, spray, or roller. Shows typical costs and bid prices for fast, medium, and slow work, including material costs per gallon; square feet covered per gallon; square feet covered per manhour; labor, material, overhead, and taxes per 100 square feet; and how much to add for profit. Includes a CD-ROM with an electronic version of the book with *National Estimator*, a stand-alone Windows™ estimating program, plus an interactive multimedia video that shows how to use the disk to compile construction cost estimates.
440 pages, 8½ x 11, $48.00. Revised annually

National Building Cost Manual

Square foot costs for residential, commercial, industrial, and farm buildings. Quickly work up a reliable budget estimate based on actual materials and design features, area, shape, wall height, number of floors, and support requirements. Includes all the important variables that can make any building unique from a cost standpoint.
240 pages, 8½ x 11, $23.00. Revised annually

CD Estimator

If your computer has Windows™ and a CD-ROM drive, CD Estimator puts at your fingertips 85,000 construction costs for new construction, remodeling, renovation & insurance repair, electrical, plumbing, HVAC and painting. Quarterly cost updates are available at no charge on the Internet. You'll also have the *National Estimator* program — a stand-alone estimating program for Windows™ that *Remodeling* magazine called a "computer wiz," and Job Cost Wizard, a program that lets you export your estimates to QuickBooks Pro for actual job costing. A 60-minute interactive video teaches you how to use this CD-ROM to estimate construction costs. And to top it off, to help you create professional-looking estimates, the disk includes over 40 construction estimating and bidding forms in a format that's perfect for nearly any Windows™ word processing or spreadsheet program. **CD Estimator is $68.50**

Fences & Retaining Walls

Everything you need to know to run a profitable business in fence and retaining wall contracting. Takes you through layout and design, construction techniques for wood, masonry, and chain link fences, gates and entries, including finishing and electrical details. How to build retaining and rock walls. How to get your business off to the right start, keep the books, and estimate accurately. The book even includes a chapter on contractor's math.
400 pages, 8½ x 11, $23.25

Contractor's Plain-English Legal Guide

For today's contractors, legal problems are like snakes in the swamp — you might not see them, but you know they're there. This book tells you where the snakes are hiding and directs you to the safe path. With the directions in this easy-to-read handbook you're less likely to need a $200-an-hour lawyer. Includes simple directions for starting your business, writing contracts that cover just about any eventuality, collecting what's owed you, filing liens, protecting yourself from unethical subcontractors, and more. For about the price of 15 minutes in a lawyer's office, you'll have a guide that will make many of those visits unnecessary. Includes a CD-ROM with blank copies of all the forms and contracts in the book.
272 pages, 8½ x 11, $49.50

Renovating & Restyling Older Homes

Any builder can turn a run-down old house into a showcase of perfection — if the customer has unlimited funds to spend. Unfortunately, most customers are on a tight budget. They usually want more improvements than they can afford — and they expect you to deliver. This book shows how to add economical improvements that can increase the property value by two, five or even ten times the cost of the remodel. Sound impossible? Here you'll find the secrets of a builder who has been putting these techniques to work on Victorian and Craftsman-style houses for twenty years. You'll see what to repair, what to replace and what to leave, so you can remodel or restyle older homes for the least amount of money and the greatest increase in value. **416 pages, 8½ x 11, $33.50**

Estimating Home Building Costs

Estimate every phase of residential construction from site costs to the profit margin you include in your bid. Shows how to keep track of manhours and make accurate labor cost estimates for footings, foundations, framing and sheathing finishes, electrical, plumbing, and more. Provides and explains sample cost estimate worksheets with complete instructions for each job phase. **320 pages, 5½ x 8½, $17.00**

Rough Framing Carpentry

If you'd like to make good money working outdoors as a framer, this is the book for you. Here you'll find shortcuts to laying out studs; speed cutting blocks, trimmers and plates by eye; quickly building and blocking rake walls; installing ceiling backing, ceiling joists, and truss joists; cutting and assembling hip trusses and California fills; arches and drop ceilings — all with production line procedures that save you time and help you make more money. Over 100 on-the-job photos of how to do it right and what can go wrong. **304 pages, 8½ x 11, $26.50**

Construction Forms & Contracts

125 forms you can copy and use — or load into your computer (from the FREE disk enclosed). Then you can customize the forms to fit your company, fill them out, and print. Loads into *Word* for Windows™, *Lotus 1-2-3*, *WordPerfect*, *Works*, or *Excel* programs. You'll find forms covering accounting, estimating, fieldwork, contracts, and general office. Each form comes with complete instructions on when to use it and how to fill it out. These forms were designed, tested and used by contractors, and will help keep your business organized, profitable and out of legal, accounting and collection troubles. Includes a CD-ROM for Windows™ and Mac. **432 pages, 8½ x 11, $41.75**

How to Succeed With Your Own Construction Business

Everything you need to start your own construction business: setting up the paperwork, finding the work, advertising, using contracts, dealing with lenders, estimating, scheduling, finding and keeping good employees, keeping the books, and coping with success. If you're considering starting your own construction business, all the knowledge, tips, and blank forms you need are here. **336 pages, 8½ x 11, $28.50**

Contractor's Guide to the Building Code Revised

This new edition was written in collaboration with the International Conference of Building Officials, writers of the code. It explains in plain English exactly what the latest edition of the Uniform Building Code requires. Based on the 1997 code, it explains the changes and what they mean for the builder. Also covers the Uniform Mechanical Code and the Uniform Plumbing Code. Shows how to design and construct residential and light commercial buildings that'll pass inspection the first time. Suggests how to work with an inspector to minimize construction costs, what common building shortcuts are likely to be cited, and where exceptions may be granted.
320 pages, 8½ x 11, $39.00

Contractor's Guide to QuickBooks Pro 2003

This user-friendly manual walks you through QuickBooks Pro's detailed setup procedure and explains step-by-step how to create a first-rate accounting system. You'll learn in days, rather than weeks, how to use QuickBooks Pro to get your contracting business organized, with simple, fast accounting procedures. On the CD included with the book you'll find a QuickBooks Pro file preconfigured for a construction company (you drag it over onto your computer and plug in your own company's data). You'll also get a complete estimating program, including a database, and a job costing program that lets you export your estimates to QuickBooks Pro. It even includes many useful construction forms to use in your business.
336 pages, 8½ x 11, $47.75

Also available: **Contractor's Guide to QuickBooks Pro 2002, $46.50**
Contractor's Guide to QuickBooks Pro 2001, $45.25

Basic Lumber Engineering for Builders

Beam and lumber requirements for many jobs aren't always clear, especially with changing building codes and lumber products. Most of the time you rely on your own "rules of thumb" when figuring spans or lumber engineering. This book can help you fill the gap between what you can find in the building code span tables and what you need to pay a certified engineer to do. With its large, clear illustrations and examples, this book shows you how to figure stresses for pre-engineered wood or wood structural members, how to calculate loads, and how to design your own girders, joists and beams. Included FREE with the book — an easy-to-use limited version of NorthBridge Software's *Wood Beam Sizing* program.
272 pages, 8½ x 11, $38.00

Basic Plumbing with Illustrations, Revised

This completely-revised edition brings this comprehensive manual fully up-to-date with all the latest plumbing codes. It is the journeyman's and apprentice's guide to installing plumbing, piping, and fixtures in residential and light commercial buildings: how to select the right materials, lay out the job and do professional-quality plumbing work, use essential tools and materials, make repairs, maintain plumbing systems, install fixtures, and add to existing systems. Includes extensive study questions at the end of each chapter, and a section with all the correct answers.
384 pages, 8½ x 11, $33.00

National Repair & Remodeling Estimator

The complete pricing guide for dwelling reconstruction costs. Reliable, specific data you can apply on every repair and remodeling job. Up-to-date material costs and labor figures based on thousands of jobs across the country. Provides recommended crew sizes; average production rates; exact material, equipment, and labor costs; a total unit cost and a total price including overhead and profit. Separate listings for high- and low-volume builders, so prices shown are specific for any size business. Estimating tips specific to repair and remodeling work to make your bids complete, realistic, and profitable. Includes a CD-ROM with an electronic version of the book with *National Estimator*, a stand-alone *Windows*™ estimating program, plus an interactive multimedia video that shows how to use the disk to compile construction cost estimates.
296 pages, 8½ x 11, $48.50. Revised annually

Markup & Profit: A Contractor's Guide

In order to succeed in a construction business, you have to be able to price your jobs to cover all labor, material and overhead expenses, and make a decent profit. The problem is knowing what markup to use. You don't want to lose jobs because you charge too much, and you don't want to work for free because you've charged too little. If you know how to calculate markup, you can apply it to your job costs to find the right sales price for your work. This book gives you tried and tested formulas, with step-by-step instructions and easy-to-follow examples, so you can easily figure the markup that's right for your business. Includes a CD-ROM with forms and checklists for your use. **320 pages, 8½ x 11, $32.50**

Craftsman Book Company
6058 Corte del Cedro
P.O. Box 6500
Carlsbad, CA 92018

☎ 24 hour order line
1-800-829-8123
Fax (760) 438-0398

Name _____
e-mail address (for order tracking and special offers) _____
Company _____
Address _____
City/State/Zip _____ ○ This is a residence
Total enclosed_____(In California add 7.25% tax)
We pay shipping when your check covers your order in full.

In A Hurry?
We accept phone orders charged to your
○ Visa, ○ MasterCard, ○ Discover or ○ American Express
Card# _____
Exp. date_____Initials _____

Tax Deductible: Treasury regulations make these references tax deductible when used in your work. Save the canceled check or charge card statement as your receipt.

Order online http://www.craftsman-book.com
Free on the Internet! Download any of Craftsman's estimating database for a 30-day free trial! www.craftsman-book.com/downloads

10-Day Money Back Guarantee

- ○ 38.00 Basic Lumber Engineering for Builders
- ○ 33.00 Basic Plumbing with Illustrations
- ○ 68.50 CD Estimator
- ○ 41.75 Construction Forms & Contracts with a CD-ROM for *Windows*™ and Macintosh.
- ○ 45.25 Contractor's Guide to QuickBooks Pro 2001
- ○ 46.50 Contractor's Guide to QuickBooks Pro 2002
- ○ 47.75 Contractor's Guide to QuickBooks Pro 2003
- ○ 39.00 Contractor's Guide to the Building Code Revised
- ○ 49.50 Contractor's Plain-English Legal Guide
- ○ 17.00 Estimating Home Building Costs
- ○ 23.25 Fences and Retaining Walls
- ○ 28.50 How to Succeed w/Your Own Construction Business
- ○ 32.50 Markup & Profit: A Contractor's Guide
- ○ 23.00 National Building Cost Manual
- ○ 47.50 National Construction Estimator with FREE *National Estimator* on a CD-ROM.
- ○ 48.00 National Painting Cost Estimator with FREE *National Estimator* on a CD-ROM.
- ○ 48.50 National Repair & Remodeling Estimator with FREE *National Estimator* on a CD-ROM.
- ○ 33.00 Painter's Handbook
- ○ 33.50 Renovating & Restyling Older Homes
- ○ 26.50 Rough Framing Carpentry
- ○ 28.50 Paint Contractor's Manual
- ○ FREE Full Color Catalog

Prices subject to change without notice

Craftsman Book Company
6058 Corte del Cedro
P.O. Box 6500
Carlsbad, CA 92018

☎ **24 hour order line**
1-800-829-8123
Fax (760) 438-0398

10-Day Money Back Guarantee

○ 38.00 Basic Lumber Engineering for Builders
○ 33.00 Basic Plumbing with Illustrations
○ 68.50 CD Estimator
○ 41.75 Construction Forms & Contracts with a CD-ROM for Windows™ and Macintosh.
○ 45.25 Contractor's Guide to QuickBooks Pro 2001
○ 46.50 Contractor's Guide to QuickBooks Pro 2002
○ 47.75 Contractor's Guide to QuickBooks Pro 2003
○ 39.00 Contractor's Guide to the Building Code Revised
○ 49.50 Contractor's Plain-English Legal Guide
○ 17.00 Estimating Home Building Costs
○ 23.25 Fences and Retaining Walls
○ 28.50 How to Succeed w/Your Own Construction Business
○ 32.50 Markup & Profit: A Contractor's Guide
○ 23.00 National Building Cost Manual
○ 47.50 National Construction Estimator with FREE *National Estimator* on a CD-ROM.
○ 48.00 National Painting Cost Estimator with FREE *National Estimator* on a CD-ROM.
○ 48.50 National Repair & Remodeling Estimator with FREE *National Estimator* on a CD-ROM.
○ 33.00 Painter's Handbook
○ 33.50 Renovating & Restyling Older Homes
○ 26.50 Rough Framing Carpentry
○ 28.50 Paint Contractor's Manual
○ FREE Full Color Catalog

Prices subject to change without notice

Name _____
e-mail address (for order tracking and special offers) _____
Company _____
Address _____
City/State/Zip _____ ○ This is a residence
Total enclosed _____ (In California add 7.25% tax)
We pay shipping when your check covers your order in full.

In A Hurry?
We accept phone orders charged to your
○ Visa, ○ MasterCard, ○ Discover or ○ American Express

Card# _____
Exp. date _____ Initials _____

Tax Deductible: Treasury regulations make these references tax deductible when used in your work. Save the canceled check or charge card statement as your receipt.

Order online http://www.craftsman-book.com
Free on the Internet! Download any of Craftsman's estimating costbooks for a 30-day free trial! www.craftsman-book.com/downloads

---

Craftsman Book Company
6058 Corte del Cedro
P.O. Box 6500
Carlsbad, CA 92018

☎ **24 hour order line**
1-800-829-8123
Fax (760) 438-0398

10-Day Money Back Guarantee

○ 38.00 Basic Lumber Engineering for Builders
○ 33.00 Basic Plumbing with Illustrations
○ 68.50 CD Estimator
○ 41.75 Construction Forms & Contracts with a CD-ROM for Windows™ and Macintosh.
○ 45.25 Contractor's Guide to QuickBooks Pro 2001
○ 46.50 Contractor's Guide to QuickBooks Pro 2002
○ 47.75 Contractor's Guide to QuickBooks Pro 2003
○ 39.00 Contractor's Guide to the Building Code Revised
○ 49.50 Contractor's Plain-English Legal Guide
○ 17.00 Estimating Home Building Costs
○ 23.25 Fences and Retaining Walls
○ 28.50 How to Succeed w/Your Own Construction Business
○ 32.50 Markup & Profit: A Contractor's Guide
○ 23.00 National Building Cost Manual
○ 47.50 National Construction Estimator with FREE *National Estimator* on a CD-ROM.
○ 48.00 National Painting Cost Estimator with FREE *National Estimator* on a CD-ROM.
○ 48.50 National Repair & Remodeling Estimator with FREE *National Estimator* on a CD-ROM.
○ 33.00 Painter's Handbook
○ 33.50 Renovating & Restyling Older Homes
○ 26.50 Rough Framing Carpentry
○ 28.50 Paint Contractor's Manual
○ FREE Full Color Catalog

Prices subject to change without notice

Name _____
e-mail address (for order tracking and special offers) _____
Company _____
Address _____
City/State/Zip _____ ○ This is a residence
Total enclosed _____ (In California add 7.25% tax)
We pay shipping when your check covers your order in full.

In A Hurry?
We accept phone orders charged to your
○ Visa, ○ MasterCard, ○ Discover or ○ American Express

Card# _____
Exp. date _____ Initials _____

Tax Deductible: Treasury regulations make these references tax deductible when used in your work. Save the canceled check or charge card statement as your receipt.

Order online http://www.craftsman-book.com
Free on the Internet! Download any of Craftsman's estimating costbooks for a 30-day free trial! www.craftsman-book.com/downloads

---

Mail This Card Today
For a Free Full Color Catalog

Over 100 books, annual cost guides and estimating software packages at your fingertips with information that can save you time and money. Here you'll find information on carpentry, contracting, estimating, remodeling, electrical work, and plumbing.

All items come with an unconditional 10-day money-back guarantee. If they don't save you money, mail them back for a full refund.

Name _____
e-mail address (for special offers) _____
Company _____
Address _____
City/State/Zip _____

Craftsman Book Company / 6058 Corte del Cedro / P.O. Box 6500 / Carlsbad, CA 92018

BUSINESS REPLY MAIL
FIRST CLASS MAIL PERMIT NO. 271 CARLSBAD, CA

POSTAGE WILL BE PAID BY ADDRESSEE

  Craftsman Book Company
6058 Corte del Cedro
P.O. Box 6500
Carlsbad, CA 92018-9974

NO POSTAGE
NECESSARY
IF MAILED
IN THE
UNITED STATES

BUSINESS REPLY MAIL
FIRST CLASS MAIL PERMIT NO. 271 CARLSBAD, CA

POSTAGE WILL BE PAID BY ADDRESSEE

 Craftsman Book Company
6058 Corte del Cedro
P.O. Box 6500
Carlsbad, CA 92018-9974

NO POSTAGE
NECESSARY
IF MAILED
IN THE
UNITED STATES

BUSINESS REPLY MAIL
FIRST CLASS MAIL PERMIT NO. 271 CARLSBAD, CA

POSTAGE WILL BE PAID BY ADDRESSEE

 Craftsman Book Company
6058 Corte del Cedro
P.O. Box 6500
Carlsbad, CA 92018-9974

NO POSTAGE
NECESSARY
IF MAILED
IN THE
UNITED STATES